KEEPING YOUR AI UNDER CONTROL

A PRAGMATIC GUIDE TO IDENTIFYING, EVALUATING, AND QUANTIFYING RISKS

Anand Tamboli

Apress®

Keeping Your AI Under Control: A Pragmatic Guide to Identifying, Evaluating, and Quantifying Risks

Anand Tamboli
New South Wales, NSW, Australia

ISBN-13 (pbk): 978-1-4842-5466-0   ISBN-13 (electronic): 978-1-4842-5467-7
https://doi.org/10.1007/978-1-4842-5467-7

Managing Director, Apress Media LLC: Welmoed Spahr
Acquisitions Editor: Shiva Ramachandran
Development Editor: Rita Fernando
Coordinating Editor: Rita Fernando

Cover designed by eStudioCalamar

Distributed to the book trade worldwide by Springer Science+Business Media New York, 233 Spring Street, 6th Floor, New York, NY 10013. Phone 1-800-SPRINGER, fax (201) 348-4505, e-mail orders-ny@springer-sbm.com, or visit www.springeronline.com. Apress Media, LLC is a California LLC and the sole member (owner) is Springer Science + Business Media Finance Inc (SSBM Finance Inc). SSBM Finance Inc is a **Delaware** corporation.

For information on translations, please e-mail rights@apress.com, or visit http://www.apress.com/rights-permissions.

Apress titles may be purchased in bulk for academic, corporate, or promotional use. eBook versions and licenses are also available for most titles. For more information, reference our Print and eBook Bulk Sales web page at http://www.apress.com/bulk-sales.

Any source code or other supplementary material referenced by the author in this book is available to readers on GitHub via the book's product page, located at www.apress.com/9781484254660. For more detailed information, please visit http://www.apress.com/source-code.

Printed on acid-free paper

*To everyone who believes that humans are
more valuable than machines!*

Early praise for "*Keeping Your AI Under Control*"

"Trusted AI and bias detection in AI are essential areas that have gained significance in recent times. Most of the AI use cases today need explainability as a critical feature. This book has excellent use cases for CIOs to assess their AI projects and their current effectiveness. In addition, the book covers an important aspect—Responsible AI, highlighted through "The Leash System" that outlines how organizations can perform a sanity check on their AI projects."

—Shalini Kapoor, Director & CTO - Watson IoT, IBM India

"In today's world of cloud services and AUTOML [automated machine learning] frameworks, the development of AI models has become quite methodical (if not easy) and is framework driven. When you have defined steps or a pipeline to approach a particular problem, it has an inherent self-correcting mechanism. Today, we have a framework for AI development and implementations from a process and technology standpoint, but we don't have a one for designing a responsible AI model with a self-correcting or feedback mechanism to learn from mistakes and be responsible for them. This book fills the gap with "The Leash System" to implement AI in an unbiased way, make it more responsible for its utility point of view, and make your models more real (if not human). AI models do not have a mind of their own, so this book will help the AI designers to design AI solutions, representing a responsible human mind."

—Rahul Kharat, Director, Head of Artificial Intelligence, Eton Solutions LP

"I had no idea AI would be such a huge game-changer when we programmed our first expert systems based application in 1996. *Keeping Your AI Under Control* not only accounts how AI is changing the world, but more importantly it gives us guidelines on how AI should be built to high ethical standards. Anand is skilled in converting complicated topics and technical jargon into a simple language; thus, the book will equally appeal to programmers and end-users.

This book helps strike a collective debate on how humanity should use a potent tool like AI towards the betterment of all."

—Amit Danglé, Vice President of Customer Success,
Saviant Consulting

Contents

About the Author

Anand Tamboli helps people to leverage emerging technologies wisely and adapt to the dynamic & disruptive future. He is a personal & business transformation specialist with distinctive traits.

As a professional speaker, Anand educates the audience on the topics of disruption, adaptability, future technologies, future of leadership, innovation, thought leadership, communication & execution, and other contemporary matters. Being a reformed futurist, he can often shed new light on a topic, which you feel has been *"done to death".*

Being a prolific writer, Anand has authored several articles to date. Anand is also the author of *Build Your Own IoT Platform* (Apress, 2019), an award-winning book on the Internet of Things.

Anand has been speaking professionally since 2009 and has spoken on a wide range of contemporary topics. When not speaking at conferences, you will find him helping professionals in personal transformation. He also coaches and mentors individuals & start-ups to maximise their value proposition.

Anand has extensively worked with several Fortune 500 multinationals such as HSBC, LG Electronics, Commonwealth Bank of Australia, and likes of them. He has thorough experience in Manufacturing, Education, Pharmaceutical, Telecommunications, IT, Banking, Insurance, and other few industries. This cross-domain experience enables him to see things with a uniquely different lens.

Anand is an intense spiritual seeker and loves to talk about core systems that shape our lives - including education, work, politics, parenting, and spirituality. Being a full spectrum thinker, he firmly believes, *"Our future is whatever we make it, so let's make it a good one!".*

Reach him at https://www.anandtamboli.com.

Acknowledgments

There is a famous saying—"*It takes a village to raise a child*"—and it is no less the case when writing a book.

Having an idea and turning it into a book is as hard as it sounds. The experience is both internally challenging and rewarding. None of this would have been possible without so many supportive people in my life.

I believe that the support always starts with our innermost circle, with people who live and breathe with you. As for me, there are three dots in my intimate circle, my wife *Jyoti*, my son *Aadi*, and daughter *Manasvi*. These three dots are extremely good at inspiring and energizing me in the most innovative ways. Throughout the journey of writing this book, they made sure that I'm motivated, inspired, and well-fed!

I am eternally grateful to *my parents*, the source of my writing genes, who always nurtured my love of books since childhood and supported my quests to their full abilities, all their life.

I appreciate help from *Gregory Bell*. Greg helped with the initial reviews of a few chapters in the book and provided with his feedback. I remember my long discussions with Greg, which always helped in crystallizing several points I make throughout this book. Thanks Greg!

An exciting 30-day coffee challenge resulted in a meeting with *Chris Stallard*, who helped me with one of the critical chapters in this book. Chris comes from a general insurance background and has extensive experience in this domain. His inputs significantly helped in clarifying the AI insurance concept, factual writing, and making that chapter possible. Thanks mate!

Compared to my previous book on Internet of Things, my approach in writing this book was quite different. It was to formulate a concept, conduct extensive research around it, then to test my idea in the market for feedback, and finally write it down in a refined format. That involved several presentations and meetings throughout the last year. In doing so, *Paulina Kabaczuk* from *Deloitte* and *Zara Crichton* and *Nathan Plummer* from *Venture Café Sydney* have been instrumental in helping me with the platform to present and speak at the events. It helped in getting useful feedback and refining the contents. Folks, thank you!

It was *Lalit Kumar* who helped with many brainy discussions in conceptualizing *The Leash System*—a practical method for responsible AI development. This system has been a foundation for the process and content explained in this book. Thanks Lalit!

Finally, to all those who have been a part of my getting there: many thanks to everyone on the *Apress* team who helped in getting this book out in the market. Special thanks to *Nikhil Karkal* for making an introduction and *Shivangi Ramchandran* for taking the proposal further. Thank you, *Rita Fernando*, for being a patient and accommodating editor.

Introduction

Future tech always brings us two things: promise and consequences. It's those consequences the *responsible AI* is all about, and if it is not, then it should be.

The sensational news and resulting hysteria about the future of artificial intelligence are everywhere. Hyperbolic representation by the media has made many people believe that they are already living in the future.

Much of our daily lives as consumers intertwine with artificial intelligence. There is no doubt that artificial intelligence is a powerful technology, and with that power comes responsibility!

AI hyperbole has given rise to "*AI solutionism*" mindset. Such that many believe that if you give them enough data, their machine learning algorithms can solve all of humanity's problems.

We already have seen the rise of a similar mindset a few years ago, which was "*there is an app for it*" mindset, and we know that it hasn't done any good in real life. Instead of supporting any progress, it endangers the value of emerging technology and sets unrealistic expectations.

AI solutionism has also led to a reckless push to use AI for anything and everything. This push is making several companies to take the *ready-fire-aim* approach, which is not only detrimental to the company's growth but also is dangerous for customers on many levels.

A better approach would be to consider suitability and applicability, apply phronesis, and do what is necessary. However, fear of missing out is leading to several missteps and is eventually creating a substantial intellectual debt that we may never be able to pay.

One of the many ways to handle this is being responsible with AI and keeping it always in your control. However, the problem with the responsible AI paradigm is that everyone knows *why* it is necessary, but no one knows *how* to achieve it.

This book aims at guiding you toward responsible AI with actionable details. The responsible AI is not just a fancy term or an abstract concept—it is to be ethical, careful, controlled, cautious, reasonable, and accountable, that is, to be responsible in designing, developing, deploying, and using AI.

Throughout the book, you will learn about the various risks involved in developing and using AI solutions. Once you are enabled to identify risks, you will be able to evaluate and quantify them. Doing this in a structured manner means your approach will be more responsible for designing and using AI.

Knowing what we don't know has significant advantages. Unfortunately, AI tech giants are continually pushing user companies in a danger zone, where companies don't know what they don't know. This push is dangerous, not only from a risk concentration perspective but also for your own business's sake. You must seek to understand what is inside the AI black box.

Applying structured methods to evaluate risks and diligently managing them can help you to understand risk exposure and mitigate it as much as possible. The book will enable you to do it and help you to remain in control.

Any effort to solve problems must cover three key aspects—doing the right thing, doing it right, and covering all the bases. The methodology explained throughout the book will help you understand these aspects in detail.

I have organized the contents of the book into three parts. In the first part, I explore the future state of AI by assessing what is to come. I not only talk about past fiascoes and lessons learned but also elaborate on risks that artificial intelligence as a technology is posing.

The second part is about risk prevention. I take you to through evaluation method for various risks, such as AI solution risks, during the deployment, and human–machine interaction related. Additionally, I also discuss a systematic approach to risk mitigation.

While prevention is always the best strategy than cure, having a better control system and mitigation plans in place can be quite useful. The third part covers the mitigation aspect. I elaborate on the use of *red teams* for AI solutions' stress testing, along with residual risk management.

Subsequently, I also talk about AI insurance that could cover you from the new unknowns. I was careful while writing the chapter where I talk about AI insurance, and I developed it after consulting with insurance industry experts vetted the content.

Not knowing what you don't know is a more significant risk than any known risks. Identifying and mitigating such risks can give you peace of mind. And even if things go wrong, you would be able to handle them, or at least your losses may be covered.

Technology is not everything; it also takes an appropriate combination of people, process, and timing to achieve the best results. The book helps you cover all these aspects.

Remember that you can develop a responsible AI specific to your use case if you know the risks and understand them better.

I not only invite you to read the book but also to try applying techniques and methods explained for your company and educate others about the same— we will need collective efforts to make AI responsible again!

Future State of AI

Assessing what is to come

Artificial Intelligence Beyond 2020

Artificial intelligence or *AI* has finally made its way to the list of mainstream technologies, and it has done so much faster than anticipated. However, its journey from this stage forward is rockier than it was for other technologies in the past. It is a common phenomenon that if you repeat a word enough number of times, it loses its meaning. This is the current state of AI.

This book is about understanding AI from a realistic point of view. It would be prudent to discuss some of the reasons and understand why AI has not been able to make its way into more than a few aspects of the business until now.

But before we do that, let us start by discussing the current state of AI solutions, implementations, and common themes that we have seen in the past couple of years. Keeping these trends in mind, we should be able to extrapolate its journey through the year 2020 and beyond. To be clear, this by no means is a crystal ball prediction. Given that technology is growing exponentially, things may happen sooner or later than anticipated!

© Anand Tamboli 2019
A. Tamboli, *Keeping Your AI Under Control*,
https://doi.org/10.1007/978-1-4842-5467-7_1

We have changed gears recently

The concept of AI has been around for centuries; however, it took off significantly during the 1950s, when Alan Turing explored the real possibilities of this concept. Due to the state of computer hardware available at that time, this work did not progress much.

When computers became more powerful in later years, they were faster and affordable and had more power in terms of storage as well as computing speed. Since then, research in AI has been growing steadily. From mere 1MB memory big-box systems to 128GB memory credit card–sized systems, the advancement in hardware has enabled technological augmentation by leaps and bounds.

However, in the past few years, there has been sudden growth in all the activities related to AI. The realization of the Internet of Things (*IoT*) and other complementary technologies such as Big Data, Cloud Computing, etc. supported this growth.

Since last year, we see less of AI theory and more of AI implementations. Without any doubt, AI technology is still in its nascent stage, but it has reached a critical mass, where research and application can happen simultaneously. We can undoubtedly say that we have changed gears.

AI has covered a lot of ground

Perhaps the most widely used and easy to deploy AI application has been the customer service online chatbot. Despite their ease of use, there are still several bots that do not work as intended, and your one intelligent question can throw them off. It is interesting to note that, although there are proven applications for structured language processing in the form of compilers, this application (online chatbot) still struggles a lot.

Image recognition is another widely used application for several use cases. Right from tweaking the photos you take on your phone to the identification of number plates of a moving vehicle, it is being used prominently. Facial recognition has reached acceptable accuracy. However, there are significant aspects affecting privacy, security, and the like. These aspects have affected universal rollout.

Email and social media are increasingly using primary forms of AI where message filtering, smart email replies (e.g., Gmail), or personalization of ad contents is being done regularly.

Tesla's predictive capabilities, self-driving features, is one such area where AI is making its mark.

Various other applications like Netflix and Pandora have implemented AI for content suggestions to the consumer based on their historic preferences.

In the healthcare sector, primarily pharma domain, neural networks are being used for speedier drug discovery.

Creating standard technical content such as product descriptions, specification sheets, and so on has also been another recent AI application.

Applicant tracking systems along with resume scanning has been in use by recruitment industry for quite some time. New applications also have some capability to scrutinize your resume and show the matching percentage with given job requirements. This capability is quite similar to showing you search results based on your search engine keyword query. It has its quirks though, but regardless, it is not going away.

Banking and insurance sectors are judging your creditworthiness based on inputs and information you provide to them. Due to their abundant data on you, as well as central databases like credit agencies and so on, these systems can make pretty close judgment calls. In some cases, their functioning does get affected due to lack of data or information, especially when privacy legislations prohibit them from accessing certain information. However, beyond helping in operations space in the finance industry, several applications have been developed for advising customers on investments and currency or stock trading, and so on. As such, AI and the finance industry are said to be a match made in heaven.

General automation in many service industries, such as the telecommunications sector, is also growing significantly. AI systems are now scanning the network and can run several tests in parallel to investigate and resolve problems as they arise.

In short, AI is already making decisions that affect your life, whether you like it or not, and has covered significant ground in recent past years.

But it is not everywhere yet

While it would be natural to think that AI has penetrated almost every single vertical or market, it is far from the truth. At best, there are only a few technology spot fires in a few select industries where AI is making its mark. Unfortunately, as always, while marketing gimmicks are at play to make everyone feel that AI has covered everything; several sections are still untouched, and that is for the right reasons.

Many image recognition systems are now better at detecting cancer or microfractures from patient's MRI or X-ray reports. Many pattern recognition systems can correlate several pathological reports and make an almost precise prediction of the health status of the patient. And yet, medical recommendations without doctor's explicit approval or signatures are not a commonplace practice. And this is good, because when there is human life at stake, systems should not make a final call, ever. Therefore, as far as the medical field is concerned, AI might only reach a status of assisted intelligence and may not be permitted (should not be allowed) to become a mainstream phenomenon at all.

Arms control is a little bit of a gray area at the moment, where granting licenses is not AI driven and the likelihood of that becoming a mainstream application is almost zero for all the right reasons. It is a different argument though that you may not need arms when you have AI easily accessible.

While companies are continually taking humans out of the customer service sector and replacing them with chatbots or automated responders that are AI driven, a human touch is becoming expensive. I already saw a startup pitch at an event, where their primary differentiation was "we provide *personal support for all your queries.*" Mostly, we see an exciting shift in terms of AI- and non-AI-based solution offering.

Self-learning applications is another area where AI is making an entry. Using customized learning, pace, and recommendations, it is becoming popular. However, as that happens, teaching, coaching, and mentoring will soon become a high-touch service and will still be in demand. Therefore, it is difficult to say, whether AI has touched this sector truly or just morphed it into something else.

Another aspect where AI has not yet touched and would not touch is live entertainment and art. These are such personalized and creative pursuits that without having a human in it, they would not have the same meaning. However, there have been a few experiments with AI creating art, but those art forms have quite a different flavor to it. AI systems can create art based on what they have been trained for. Several of those are mainly geometrical and systematic shapes or pictures, nothing that a human would necessarily draw with a slightly acceptable and natural imbalance in it. The real authorship of the work of art cannot be yet bestowed to an artificial system.

Creativity is some part process and some part randomness. Being the exact opposite of the rule-based process, AI will not be able to contribute directly to the creative industry any time soon.

How do end users see AI?

As far as end users of AI technology are concerned, there is high-level fear, uncertainty, and doubt (FUD) among the majority.

The sheer duality of this technology is a significant concern. AI is a powerful tool, and just like any other tool, humans can use it for good or bad things. However, since we aren't actively talking about how to handle potential misuse of AI, this has remained as a growing concern.

Another reason for having a skeptical outlook toward the AI is a plausible fear of job losses. If there are massive numbers of people losing jobs without an alternative system in place, it would be undoubtedly dangerous, and this can create chaos. But then again, if you think about it deeply, you will realize that it is not losing a job that concerns many. What people usually worry about is having nothing better to do when their mainstream work is disrupted.

Unfortunately, the majority of AI implementation projects do not address this issue up front. Instead, it is done as an afterthought. This is perhaps the most substantial reason for being skeptical of AI.

At a very superficial level, many of us do appreciate ease and convenience these AI solutions are providing. However, our comfort soon erodes as these solutions start to increase their scope and touch critical areas of our life, such as banking, social benefits, security, healthcare, and others.

Bias and racism have been front-runners in the list of reasons for the distrust in AI as such. People also fear that AI may show blatant disregard for human control. This, however, does not have any precedence, but it is practically possible, and hence, it is a legitimate concern.

Errors at scale is not a widely known issue, but those who have been victims of this problem in the past see this as one of the significant concerns when using AI in daily life. Imagine when a public AI system cancels credit cards of thousands of people because of some error. The scale of chaos this may cause is the main reason for this concern.

As a general observation, I have seen that end users of AI are comfortable for as long as the applications are not touching or affecting core life matters. They are comfortable in areas of entertainment and luxury, but not so much when critical aspects of life are in the hands of an AI such as finances, health, security, jobs, driving, and others alike.

What are business users thinking?

Now, despite the commonplace mixed feelings and heightened expectations from AI, the business world still has some ability to see AI in a relatively balanced manner. People from a wide range of industries even agree that AI is tricky to deploy, and it could suck a lot of money and time before becoming useful. It can be costly, and the initial payout can be quite modest. The overall payback period for any AI project as such hasn't been attractive per se, and in many cases, it is hard to establish objectively.

Several experts, including myself, find it unsettling that some vendors are pushing AI systems even before they figure out the purpose and claim to know what problem it will solve. Some businesses discourage and loathe this approach and are taking a prudent approach, but that is only a small minority of them.

It is one thing to see breakthroughs in gaming AI such as in the game of Go and Chess or having devices that turn on music at your voice command. It is another thing to use AI and make step changes in the businesses, especially organisations that are not fundamentally digital.

When it comes down to improving and changing how business gets done, AI and other tools form only small cogs of a giant wheel. Changes that bring on company-wide repercussions are a different ball game.

The change management aspect hasn't been easy to handle in the past, and it is not going to change in the future either. Several experts from various domains of the business are needed to be involved for any significant change to occur, and they have to be the best ones if we are looking for effective outcomes, which essentially means that you have to pull your best people from routine business work and let them focus on AI implementation, which is a difficult proposition for business of any size.

What's to come?

Technology has come a long way over the last 20 odd years. Several technological advances in hardware and software alike have changed the way we do things and experience our life now.

Twenty years ago, computer single core CPU would run at 225 MHz. Now the same sized chip has eight core CPU which runs at 2.4 GHz. If you compare hand phones (not smartphones) that were released during the last decade, with their revised versions now, the difference would be enormous. Several factors of comparison did not even exist back then.

The trend of technological innovation has always been heading upward and rightly so.

One of the most widely discussed AI technologies today is autonomous vehicles. These will have a profound impact on how we think about transportation and would bring a significant positive change to future and current cities.

Especially for people who need assistance, such as the elderly, children, disabled people, and patients, this could mean a lot. Supply chain industry, as well as everything that depends on physical distribution network (post, shopping, etc.), would see a considerable shift in operations.

AI and other emerging technologies, besides bringing efficiencies, are also bringing new possibilities. These new possibilities are creating new business models and opportunities that we have not seen before. This will continue to happen in the future as we progress.

Most of the tasks in our daily lives that depend upon our best estimates or guesswork would also see a significant shift due to the abundance of data with regard almost anything.

As we have access to more data, the need for devices that can process this data on edge will increase and will be a key driver in maintaining this progression.

One of the significant drivers in all these technological advances has always been the democratization of the resources. Whether it's the Internet revolution, open source hardware and software revolution, or anything else, as AI technology becomes a part of our daily lives, we will see more of this democratization happening. This will be a crucial factor and will keep boosting progress.

As of now, most AI applications follow a supervised learning approach. In years to come, we will start seeing more and more of unsupervised learning that will keep systems updated continuously. However, this will have one significant barrier to cross, which is the **trust factor**. Unless this *trust factor* improves, supervision will remain a necessity.

There is no accepted or standard definition of good AI as such. In my view, however, good AI will be one that can guide users to understand various options, explain the trade-offs among multiple possible choices, and then help to make those decisions. Good AI will always honor the final decision made by the human.

On the consumer front, several virtual support tools will increase and will become mainstream. It will be almost expected to come across these bots first before talking to any human at all. However, only the businesses that demonstrate a customer-centric approach will thrive in these scenarios, while others will struggle to adapt to the right technology.

And, most importantly, *"What do you want to do when you grow up?"* will soon become an obsolete question. AI will change the job market entirely as we will see growing requirements for soft skills since most of the hard skills will be automated.

A balanced approach

Regardless of how the recent or long-term future with AI looks like, there are a few points that we must understand and accept in their entirety. Most of these points align with OECD's AI principles[1] that were released in early 2019.

---

[1] www.oecd.org/going-digital/ai/principles/

AI systems should benefit humans, overall ecosystem on the planet, and the planet itself, by driving inclusive growth, sustainable development, and well-being of all.

These systems must always be designed such that they respect and follow the rule of law and rights of the ecosystem (humans, animals, etc.). They should also respect the general human value system and diversity it exhibits. More importantly, there must be appropriate safeguards in the system such that humans are always in the loop when necessary or can intervene if they (humans) feel the need, regardless of the necessity. After all, fair and just society should be the goal of any advancement that we bring.

Creators of AI systems should always demonstrate transparency and responsible disclosure about the functionality and methodology of the system. It is essential that people involved and affected by this do understand how outcomes are derived and, if needed, should be able to challenge them.

Any AI system should not cause harm to the users or general living beings as such and must always function in a robust, secure, and safe way throughout its life cycle. Creators and managers of these systems have the responsibility to assess continually and manage any risks in this regard.

And most importantly, on the accountability front, anyone creating, developing, deploying, operating, or managing AI systems must be held accountable for the system's functioning and outcomes at all times. Accountability can drive positive behaviors and thereby can potentially ensure that all the preceding general principles have adhered.

There is a general feeling that overregulation limits innovation and advancement. However, in my view, let us not race to be the first, let us strive to be better! Being fast and first by compromising on ethics and quality is certainly not an acceptable approach by any means.

I don't think that in the next ten years or so we will have robots controlling humans. However, technology consuming us, our time, feelings, and mindfulness is very much a reality even today, and it is getting worse. Just one wrong turn in this fast lane is what it will take to cause regression for society. Let us work toward keeping the technology, AI or otherwise, in our control, always!

■ Day by day, the technology is becoming autonomous and exponentially smarter; we humans must find ways to catch up rather quickly, lest being subjugated.

Learning Lessons from Past Fiascoes

Whenever we are set to improve something, what is the first thing we do? If you are thinking of assessing the past, you are correct. It is the most convenient and sure-shot method of understanding improvement areas and fixes that are required for better future outcomes.

This should be no different for AI implementations as well. Automation and primitive forms of AI have been in trial and operation for many years. Some have worked as expected and some not so much. I have chosen a few use cases that have failed or have not shown full expected benefits.

These case studies form a reasonable basis for understanding weak spots to focus on and thereby helping to establish a baseline for risk profiling. Some standard parameters will also emerge in the form of a repetitive pattern, which will give an insight into common root causes of these fiascoes.

By no means, these use cases are a complete representation of the AI solution landscape, but they represent a few common aspects, which we will discuss subsequently. But first, the fiascoes.

© Anand Tamboli 2019
A. Tamboli, *Keeping Your AI Under Control*,
https://doi.org/10.1007/978-1-4842-5467-7_2

When Microsoft's chatbot went Nazi on social network

On March 23, 2016, Microsoft Corporation released an artificially intelligent chatbot, named "Tay." Named after the acronym "Thinking About You," Tay was developed to help better understand how AI would interact with human users online.

Subsequently, Tay was programmed to ingest tweets and learn from them to communicate by responding and interacting with those users. Mainly, the target audience was American young adults. I do not think the outcome would have been any different if the target audience was any different though. However, this attempted experiment only lasted 24 hours before Tay had to be taken offline for publishing extreme and offensive racist as well as sexist tweets.

Tay was a classic example of a system that was vulnerable to a burning and pertinent issue in the data science world—"garbage in, garbage out." Tay was developed to learn from active real-time conversations on Twitter, but it did not have an ability to filter offensive inputs or bigoted comments in the process. Perhaps that ability was there but wasn't enough for the purpose. Ultimately, Tay learned from user responses and reflected the same kind of emotion or thinking. Since most of the tweets were abusive, racist, and sexist, Tay's responses followed the pattern.

The second attempt to release Tay by Microsoft didn't go too well either, and the bot was taken down soon after the second release. It hasn't been up online since then.

When looking closely, I see that although Tay wasn't explicitly programmed for discrimination of any kind, it is safe to assume that its learning data did not have any discriminating characteristics either. However, the feedback loop from which Tay was supposed to learn seemed to have a flaw.

When Amazon's same-day delivery caused a racial disparity

In early 2016, Amazon rolled out same-day delivery to its Prime program subscribers, but only for a select group of American cities. That too was only for a select group of neighborhoods, potentially, where the concentration of Prime subscribers was large enough to justify operational costs. However, it sooner became an issue when people realized that predominantly nonwhite neighborhoods were primarily excluded from this offering. Customers from these areas weren't happy as they believed to be excluded from same-day delivery and this further marginalized them while already facing the impact of bias and discrimination.

When this issue was highlighted in several forums and the media, Craig Berman, Amazon's VP for Global Communications, asserted, "When it comes to same-day delivery, our goal is to serve as many people as we can." He continued, "Demographics play no role in it. Zero."[1] From Amazon's standpoint, this seemed to be a logical approach from the cost and efficiency perspective, such that they would prioritize areas with most existing paying members over the others.

A solely data-driven approach that looked at numbers instead of people did reinforce long-entrenched inequality in access to their services. For people who were inadvertently excluded, the fact that it was not deliberate did not make much practical difference. The fact that racial demographics tend to correlate with location or zip code did result in indirect discrimination.

Apparently, in scenarios, when zip codes are considered as one of the data points for any decision-making, there will always be some inherent bias. This is because zip codes usually represent tightly knit communities of one kind. Perhaps better sampling of input data could help to a certain extent in minimizing this risk, but it would be far from being absolute zero.

When Uber's autonomous car killed a pedestrian

In 2018, an Arizona pedestrian was killed by an automated vehicle owned by Uber.[2] A preliminary report released by the National Transportation Safety Board (NTSB) in response to the incident stated that there was a human, present in the automated vehicle, but the human was not in control of the vehicle when the collision occurred.

Accidents usually imply several things going wrong at once and hence are difficult to investigate as well as ascertain the final accountability and root cause. Various reasons contributed to this accident—poor visibility of the pedestrian, inadequate safety systems of the autonomous car, and lack of oversight by the human assistant.

The legal matter was eventually settled out of courts, and further details were not released. However, the issues around liability present significant complexity due to multiple parties and actors involved. In this occasion, the vehicle was operated by Uber but was under the supervision of a human driver who was Uber employee, and it was operated autonomously using components and systems designed by various other tech companies.

---

[1] David Ingold and Spencer Soper, "Amazon Doesn't Consider the Race of Its Customers. Should It?," *Bloomberg*, April 21, 2016, www.bloomberg.com/graphics/2016-amazon-same-day/.
[2] www.theverge.com/2018/3/28/17174636/uber-self-driving-crash-fatal-arizona-update

Responsibility attribution with an AI system poses a significant dilemma, and unfortunately, there are no set standard guidelines, which could make it more transparent.

On occasions like this, where human lives are directly affected, clarity is undoubtedly needed along with a universal framework to help in a standardized approach to deal with such matters.

Although there was no final and clear legal verdict as such, the fact that ambiguity exists is a significant concern. When there are several actors at play, human as well as nonhuman, direct as well as indirect, ascertaining accountability, root cause, and liability will be the key for clarity.

When the light bulb DoS attacked an entire smart home

This is a story from 2009, when Raul Rojas, a computer science professor at the Free University of Berlin, built one of Germany's first smart homes.[3]

Everything in his house was connected to the Internet so that lights, music, television, heating, and cooling could all be turned on and off from afar. Even the stove, oven, and microwave could be turned off with Rojas's computer.

Now the challenge with his setup was ubiquitous one; there were several manufacturers and protocols at play. To help address this, Rojas designed the whole system such that all of his smart devices were connected to one hub. The hub was mainly a coordinator of all the communications among devices as well as through the Internet.

A few years later of his installation, during 2013, his smart home gave up and stopped responding to his commands. In computer parlance, system hanged up or froze. When Rojas investigated, it turned out that there was one single culprit that caused this problem, a connected light bulb!

He found that his light fixture burned out and was trying to tell the hub that it needed attention. However, when doing this, it was continuously sending requests to the hub, which overloaded the network and eventually caused it to freeze. In other words, it was causing a denial of service (DoS) to the rest of the smart home devices by standing in their way of communication with the hub. When Rojas change the bulb, the problem was fixed promptly.

---

[3] https://splinternews.com/this-guys-light-bulb-performed-a-dos-attack-on-his-enti-1793846000

This, however, highlights a few potential problems in smart homes and otherwise so-called autonomous systems. When things go wrong and are out of hands, how an end user can take over the control and put things back to their place? Let alone controlling, how an average end user can even investigate the matter and figure out what is wrong in the first place? Rojas did not design for a bypass mechanism where he could commandeer the whole system. Maybe because it was an integration of several heterogeneous systems or perhaps he didn't think of it!

When an Australian telco wasted millions of dollars

In early 2018, an Australian telco (telecommunications company) bit the bullet and rolled out an automation bot (auto-bot) for its incident handling process. Although the telco expected to reap the benefits from the beginning of its implementation and save on more than 25% of the operational costs, the plan backfired.

This auto-bot was designed to intercept all of the network incidents, 100% of them, and then follow a series of checks based on the problem statement selected by the users. The auto-bot was programmed to take one of the three actions based on the tests it would perform. It would remotely resolve the incident by fixing the issue programmatically, or it would assume that a technician's visit is required to customer premises. And accordingly it would send someone directly, or if none of that was apparent, it would present the case to the human operator for further investigation and decision.

This approach was sound and seemed utterly logical in the first place. However, after the commencement of auto-bot operation, within a few weeks, the telco realized that this auto-bot was sending an awful lot of technicians in the field than before when humans handled the entire operation. Sending out technicians for the field visit was a costly affair and was always the last choice for fixing an issue. However, auto-bot maximized on that.

Upon investigation, the team figured out that there were a few incident scenarios that a human operator could understand (and invariably join the dots) but were not clear enough to be programmed for auto-bot. In all such cases, a human operator would have taken a different decision than the auto-bot.

The primary issue though was, despite finding out the flaw in logic, the automation team was not able to turn off the auto-bot (much like what Microsoft did with Tay). The auto-bot was implemented in *all or nothing* manner, and it was sitting right in the middle of user and operator interface. Which essentially meant either all the incidents would go through the auto-bot (and get wrongly handled often) or none would go through the auto-bot and thus would get manually handled. Now, the only issue was that the telco was not ready to handle such a workload.

Eventually, the telco set up another project to fix the auto-bot while it was in operation and wasted millions of dollars in the process. The money was spent (or wasted) on two fronts, one for continuing the service with not so intelligent auto-bot and another for running massive fixing project that lasted for more than a year.

As the endowment effect kicked in, the telco had no plans to go back and fix the problem from its roots, but instead kept ploughing through and wasting an enormous amount of money. A critical question remains—who is eventually paying for this?

In retrospection, this implementation did go wrong on several different levels, right from system design to its implementation and fixing of the problems. But the first and foremost question that emanates is why there was no plan B, a kill switch of some sort to stop this auto-bot. It appeared that the auto-bot development and rollout were not thoroughly tested for all the potential scenarios and thus lacked testing rigor that could have identified problems early on. While the time required to fix the situation was too long, detecting the failure of auto-bot took considerably longer.

These fiascoes have something in common

Whether it is Tay the chatbot by Microsoft or Uber's autonomous car or the auto-bot of Australian telco, all of these have something in common that failed. And it is not the technology itself!

In all these scenarios, either the creator of AI or businesses that deployed the AI has not been careful enough. The fundamental tenet of handling something as powerful as automation or primitive AI wasn't followed responsibly. We often say, "With great power comes great responsibility." And yet, in all these cases, responsible design or deployment did not happen or did not occur in full spirit.

This responsible behavior is not only needed in deployment and use of AI but also required equally through all the stages, right from conception to design, testing to implementation, and ongoing management and governance.

Almost all the five cases we discussed have had a certain level of weaknesses in the solution conception stage, and this directly seeped into their development.

In the cases of Tay chatbot, Australian telco, and Uber autonomous car, emphasis on solution quality wasn't enough. There might have been a few testing routines, just enough to meet the requirements of IT development frameworks, but not enough to meet the AI development framework, which doesn't exist!

Even in the case of Amazon, the skill set of decision-makers could be questioned. Although it appeared to be an inadvertent decision to roll out same-day deliveries to only a few localities, thoughtful consideration could have raised these questions in reflection. Same goes with Australian telco case or light bulb situation, where creators lacked thoughtful design of the solution.

Key lessons

While there are several use cases of AI to learn from, I specifically chose these five, which are indicative of a common issue with AI design, development, deployment, and governance; and that is a thoughtful or responsible approach.

A few key learnings that have emerged are as follows:

- *Data governance is essential* from ethical AI's point of view; therefore, the creators of AI need to ensure they have robust data governance foundations, or their AI applications risk being fed with inappropriate data and breaching several laws.

- Narrow AI is all about the relation between input and outputs. You provide input X and get output Y, or there is input X to do output Y. Either way, the nature of input affects output such that indiscriminate input can lead to adverse outcomes. And this is just one good reason why *rigorous testing is so important*. We must note that in the case of AI systems, general IT system testing mechanisms are usually not enough.

- Automated decisions are suitable when there is a large volume of decisions to be made, and the criteria are relatively uniform and uncontested. When discretion and exceptions are required, automated systems should be only used as a tool to assist humans—or don't use them at all. There are still several applications and use cases, which are not defined as clearly as a game of chess.

- *Humans must always be kept in the loop*, no matter what, whether it is during the design phase of AI or testing or deployment. Majority of the AI systems are still in infancy, and all of them still need a responsible adult to be in charge. And most importantly ensuring enough human resources are available to handle the likely workload is always a good idea. How to estimate a reasonable workload is an important question, though.

- And most importantly, *a transparent chain of account-ability*—if the answer to a question "Who is responsible for the decision made by this AI?" does not yield single person's name, then this gap needs to be fixed.

- For customer-facing AI systems, *having a good experience all times is crucial*. If customers have a terrible experience, they would lose the trust and eventually render your AI solution useless.

- Fairness, lack of bias, interpretability, explainability, transparency, repeatability, and robustness are a few critical aspects that are some of *the must-have characteristics for a trustworthy AI* solution.

- As AI systems become more powerful, managing risk is going to be even more critical from a *good governance and risk management* point of view. Having this governance in place is not only an umbrella requirement for the industry but also is a good idea for every business to have in-house.

- The *ethical aspect must always be upheld,* and just because something can be done, should not be done. An AI that negatively affects employees, customers, or the general public, directly or indirectly, serves no good and should not exist.

Failure is not mere failure. It is instructive. The person who really thinks learns quite as much from his failures as from his successes.

—John Dewey, *American philosopher*

Understanding AI Risks and Its Impacts

In recent past years, we have seen dramatic improvements in AI, thanks to technological advances. We would continue taking longer strides with even more developments and substantial progress in the coming years. However, while this happens, we are unknowingly creating various risks to our socioeconomic structure, civilization in general, and some extent for human species.

Even if species-level risks are not evident yet, other two, socioeconomic- and civilization-level risks, are significant enough and cannot be ignored. Additionally, there are also several business-level risks that could affect business metrics adversely.

For now, let us talk about general outcome risks that can have a significant impact on critical social, civil, and business aspects. And, to have these risks must be managed. So, it is vital that we identify them first, acknowledge them as risks instead of discarding them, and then work toward mitigating those risks as we progress.

© Anand Tamboli 2019
A. Tamboli, *Keeping Your AI Under Control,*
https://doi.org/10.1007/978-1-4842-5467-7_3

Musk, Hawking, Gates—they all said it already!

Many notable individuals in the field across the globe have been warning and voicing their concerns for the past several years. A significant part of these concerns has been related to superintelligence per se. However, they also indicated speed and approach to be one of the primary drivers of these risks.

In 2014, Stephen Hawking, who was one of Britain's preeminent scientists, said that efforts to create thinking machines pose a threat to our very existence. He told the BBC[1] that the development of full artificial intelligence could spell the end of the human race. He was responding to a question about a revamp of the technology he used to communicate, which involved a basic form of AI.

On similar lines, Microsoft founder Bill Gates, in 2015,[2] echoed these concerns, "I am in the camp that is concerned about superintelligence. First, the machines will do a lot of jobs for us and not be superintelligent. That should be positive if we manage it well. A few decades after that, though, the intelligence is strong enough to be a concern."

During SXSW 2018,[3] Elon Musk said, "The biggest issue I see with so-called AI experts is that they think they know more than they do, and they think they are smarter than they actually are. This tends to plague smart people. They define themselves by their intelligence, and they don't like the idea that a machine could be way smarter than them, so they discount the idea— which is fundamentally flawed." As he continued to explain, "I am not really all that worried about the short-term stuff. Narrow AI is not a species-level risk. It will result in dislocation, in lost jobs, and better weaponry and that kind of thing, but it is not a fundamental species-level risk, whereas digital superintelligence is."

As you would have noticed, Musk has pointed toward general AI, which hasn't been a reality yet. As several experts in the field of AI would echo, it will not be a reality any sooner. However, while most of them are putting 25- to 30-year time frame for general AI to take shape, I think it would happen sooner.

Now, the question is, why is it that these famous people are concerned about AI while the rest of the world is thinking the opposite? I would call it the problem of hypocognition!

---

[1] www.bbc.com/news/technology-30290540
[2] www.bbc.com/news/31047780
[3] https://youtu.be/kzlUyrccbos

We are suffering from hypocognition

Let's first understand the term hypocognition. In cognitive linguistics, it means missing and not being able to communicate cognitive and linguistic representations due to lack of words for particular concepts. In short, there are no words to express the feelings, object, category, idea, or a notion.

An American psychiatrist and anthropologist Robert Levy first coined the word "hypocognition" in 1973.

More generally, experts often overuse their expertise, for example, cardiologist diagnoses a heart problem when the actual problem is something else.[4]

Hypocognition is about the absence of things. It is hard to recognize precisely because it is invisible. It is the same thing we are experiencing about the risks of AI. We haven't thought deeply about them, yet!

When we want to choose one door over the other, we ideally want to know what is on the other side. Based on what is there, we select the door. When we had to choose between the atomic bomb and no bomb, we chose the first option, because we did not know what was there on the other side of this choice. The same may happen with each AI system. And we humans would have to choose—the only problem is, this time, some people are telling us what is waiting on the other side for us. These people are AI's blind proponents. My question is, "should we believe them blindly?"

More so, we are not considering modular advancements that could soon build up this risk multifold. If you have experience in module developments, such as programming or using building blocks of some kind, it would be more apparent. Making the first few building blocks takes a lot of time and efforts; however, as we build them, the next build—using these blocks— becomes more straightforward, more comfortable, and faster. And within no time, the overall build becomes a way to faster. With AI developments, it is the same case. So far, several significant building blocks are being developed, and when we would stitch them together soon, the capability of AI would increase by many folds. And, this is the primary concern I believe Musk, Gates, and Hawkins would have seen.

What is perhaps more concerning is the risks we have not thought of yet. If you have had asked people in the 18th century, what are the significant risks to the civilization were, they wouldn't have said nukes or guns. We are probably in the same situation as them, and, we don't know what we don't know!

---

[4] Kaidi Wu and David Dunning, "Unknown Unknowns: The Problem of Hypocognition," *Scientific American*, August 9, 2018, https://blogs.scientificamerican.com/observations/unknown-unknowns-the-problem-of-hypocognition/.

There are several risks that this AI movement is bringing with it. We may not be able to avoid all of them, but we can understand them so we can address them.

Many AI systems are as good as junk

We are so gung ho about new technologies, including AI that we have almost stopped demanding the right quality solutions. We are so fond of this newness that we are ignoring flaws in new technologies.

This, in turn, is encouraging subpar solutions day by day, and we are sitting on a pile of junk, which is growing faster than ever. The disturbing part of that is those who are making subpar solutions, keep telling that the machines would be wiser soon while shrugging off their responsibilities.

The question however is, why should we challenge junk AI systems?

Merely pushing for more quantity (of tech solutions) without paying attention to quality (of the solutions) adds a lot of garbage out in the field. This garbage, when out there in the market, easily creates disillusionment and in turn adds to the resistance for healthy adoption.

This approach (quantity vs. quality) also encourages the corrupt practices by creators of AI at customer's expenses, where subpar solutions are pushed down customers' throats by leveraging their ignorance and an urge to give in to the fear of missing out (also called as FOMO).

This is a common scenario with several technologies that are going through the hype cycle, and AI is going through the same.

The problem though is that flaws in subpar techs do not surface until it is too late! In many of the cases, the damage is already done and would be irreversible. Unlike many other technologies, where things work or do not work, there is a significant gray area which can change its shades over a period. Moreover, if we are not clear in which direction would that be, we end up creating a junk. This junk essentially means someone somewhere has wasted a lot of money in creating and nurturing these solutions. It also indicates that several customers may have suffered or have had negative experiences, courtesy junk solutions.

This is why we not only need to challenge the quality of every solution but also improvise our approach toward the emerging technologies in general.

Most of the times, technology solutions go through several phases. The first phase usually is always minimum viable product (MVP). And it is a good practice to create an MVP solution, validate it, and test it thoroughly before going ahead with subsequent deployment.

Treating MVP as MVP and nothing more is going to be the key in avoiding junk solutions making their way to the market. Dealing with these infant solutions with utmost care and skepticism will ensure that we are not rushing toward productizing or releasing them lest creating another solution that may have a potential flaw to disrupt healthy life.

There is no balanced scorecard for AI

Not having a balanced scorecard suffered sales departments for several decades and instilled predatory sales practices. This happened because one of the goals for sales' team was to maximize sales numbers. This, however, lacked some of the finer details and did not specify acceptable methods of sale to achieve those goals.

Contention and racing or conflicting objectives are often the cause of friction within various teams and people. With AI in the picture, there is a matter of scale. That is, AI-powered solutions are known for their efficiency and effectiveness at a massive level. Therefore, misalignment between our goals and machine's goals could be dangerous for humans. It is easier to course correct a team of humans, doing that with machines could be a very tricky and arduous task.

For example, if you command to your autonomous car, "Take me to the hospital as quickly as possible," it may have terrible consequences. Your car may take you to the hospital quickly as you asked but may leave behind a trail of several accidents. Without you specifying that the rules of the road must be followed and no humans should be harmed, and no dangerous turns should be taken, and several other sets of humanity rules, your car would turn into a 2-tonne weaponized metal block.

In short, the AI might use destructive, unethical, or unacceptable methods for achieving its goals, if we fail to align our goals with it entirely. And, achieving that level of alignment is quite tricky. Without having any balanced approach like a scorecard, this may not be achievable.

AI knows "how" but not "why"

Artificially intelligent systems are usually excellent in performing any given task efficiently, effectively, and, at scale, consistently. However, knowing *how* to perform a specific task sometimes is not the only thing one needs to know. Understanding *why*, the purpose of the job, is equally essential since it can give a valuable context about the task itself. Understanding this context is helpful not only for performing the task but also improvising as and when required and ensuring that all the relevant bases are covered.

However, with the AI, since it does not know underlying data and its premise, the bias will be inherent. Moreover, since AI systems do not understand the context of the task, and solely rely on their training data, they are not dependable. The reliability of their outcomes could be risked if the input data is biased, incomplete, or of poor quality.

If we overestimate capabilities of AI and rely only on the knowledge of *how*, without knowing *why*, we will be risking the outcomes.

"We don't want to accept arbitrary decisions by entities, people, or AIs that we don't understand,"[5] says Uber AI researcher Jason Yosinkski. "In order for machine learning models to be accepted by society, we're going to need to know why they're making the decisions they're making."

Not knowing *why* can sometimes become a source of the bias in AI systems.

When AI systems only rely on the data, it becomes impossible for the users or the decision-makers to understand underlying reasons. For example, if your car insurance company tells you that you are not insurable because the system says so, and the system says so because it has data that says you were caught on camera talking on your phone while driving, would anyone even know *why* you were on the phone while driving? Could it be the case of an emergency where you had to break the rule? Only by understanding *why* it would make more sense in such cases. But it is quite possible that your insurance company may not even ask you about this and decline the cover.

When AI systems do not know why, there is always going to be a lurking risk of discrimination, bias, or an illogical outcome.

AI does not have a conscience

The conscience is the leading part of our spirit. It is a moral sense of right and wrong that is viewed to be acting as a guide to one's behavior. As Wikipedia elaborates, conscience is a cognitive process that elicits emotion and rational associations based on an individual's moral philosophy or value system.

Our ways of the world are still very much a gray area—we cannot straightforwardly explain a lot of things with a repeatable logic. This makes human behavior least replicable. However, with such a complex (human) operation, conscience plays an essential role in binding us toward a meaningful goal at all the time. It is what makes us rational in our actions and decision-making.

---

[5] https://qz.com/1146753/ai-is-now-so-complex-its-creators-cant-trust-why-it-makes-decisions/

However, when AI systems mimic humans in some way, whether it is via narrow AI or relatively broader AI or a general AI (in the future), certain human qualities are a must-have. Without which it will not be a good idea to let the AI systems roam around at free will. The human must always be kept in a loop, lest things go astray.

There are scenarios where conscience may have no role. Classic repetitive actions are an excellent example of one such situation. However, if there is even the slightest impact on end users, customers, employees, or society, human control is necessary. An AI system may take a decision but should not be allowed to act on it without confirmation from its responsible master.

It is never a good idea to believe that the computer is always right. Do not believe in AI blindly—when in doubt, prefer human decisions. Yes, we humans are flawed, yet, there are several human qualities that AI will never have or be able to replicate. Qualities and characteristics such as conscience, empathy, emotional intelligence, leadership, love, hate, and several others make us human and differentiate us from the rest of the species.

AI makes decisions based on how it has been taught to do so. Who taught it? How? With what inputs and when? All of these questions are important. Seek transparency and interpretability of AI decisions. If you cannot get hold of them, be very skeptical, you might be entering into a rabbit hole.

Whenever you are using or planning to implement AI systems in your business or life, think about the consequences. No matter how well-designed AI system is, it inherently carries certain risks at various levels. Estimate these risks objectively and then plan for implications; let me know if you need any help.

Since AI systems do not have a conscience, we must use our conscience and be prudent in utilizing them appropriately.

Losing jobs is not a real problem

One of the hot-button topics being considered as a risk from AI is job losses. However, in my view, this is mostly symptomatic and does not paint an accurate picture. We can very well call it a myth for the same reason.

A few years ago, I was working with a banking sector client, and our goal was to save roughly 1.5 hours per day per sales team member. That is whopping 7.5 hours or one working day per week, and with the team size circa 75 members, it was a significant monetary benefit. Who wouldn't have wanted that type of productivity increase?

However, we still faced significant and constant resistance, until the day when it dawned upon us, and we figured out the potential reason behind it.

We understood that the resistance was mainly due to "having nothing valuable to do if there is extra time at hand." It became crystal clear that just creating a void of 1.5 hours daily was not going to work. It was also our responsibility to fill the void that we were set to create with something useful with something that the sales team members would prefer to do if they had extra time at hand.

■ Not having anything better to do when there is extra time at hand makes people uncomfortable.

This is a quite common scenario with any new change that is set to take *something* away from us. We become uncomfortable! Not because this *something* is going away, but because if it is taken away, we don't know what we would do—unless we find *something else*, which is either same or relatively better to fill this void. If the new thing is not the same or better, but instead relatively not so good, we don't like that change, no one would.

And most of the times, we cannot fill the gap with the new activity after the fact; it is crucial that we plan it well in advance. The benefit of planning for this substitution in advance is that people respond naturally—they can let go of the first activity and create time for the new (and better) one.

I think this is the key to minimize the risk of change resistance in AI projects or any other new technologies that result in significant displacements, but losing jobs is not the real problem.

The risk of the right AI in wrong hands

I always insist that tools and systems are not hurtful; people using them are! Notably, the interface between humans and AI systems is one of the risky areas. The users of the AI system could exhibit a severe challenge by not behaving responsibly when using them. A classic example would be Uber autonomous car accident (briefly discussed in Chapter 2), where lack of operators' attention was supposedly the root cause.

Not just during the use of the systems, these interaction risks are prevalent in preliminary stages of AI system training too. Coding errors, lapses in proper data management, incorrect model trainings, and so on can easily compromise the AI system's critical characteristics such as bias, security, compliance, and the like.

Not to forget risks from disgruntled personnel getting hands on these systems and being able to corrupt systems' functioning in some way or the other.

Weapon systems equipped with AI are the most vulnerable to the *right AI in wrong hand* problems and therefore have the greatest of risks.

Again, we are not talking about AI systems *turning evil*. Instead, the concern is that a powerful AI system could be misused by someone. The possibility of AI systems being used to overpower others by some group or a country is one of the real significant risks.

Overall, the risk of the right AI in wrong hands is one of the critical challenges and warrants substantial attention to avoid it.

The gray matter—use it or lose it

Perhaps the most significant risk to humanity in my view is loss of brain power! One of the reasons why humans are in charge and not chimpanzees is the matter of intelligence, despite humans being physically weakest of most of the species. Then, what happens when we have created something more intelligent than ourselves? What is our differentiator? What USP would we have then?

Extending AI and automation, beyond logical limits, could potentially alter our perception of what humans can do. For example, several decades ago, doing a math calculation by hand on a piece of paper was highly appreciated skill for a human being. Now, with the advent of calculators and computers, we don't see much value in that (skill) anymore.

We still value human interaction, communication skills, emotional intelligence, and several other qualities in humans. What happens when an AI app takes over and starts to book appointments for us? Last year Google showcased an application where haircut appointment was booked by the app, all by itself. What happened to the opportunity to communicate with a fellow human? Are we becoming so busy to spend a couple of minutes and talk to another person? What happened to AI doing mundane tasks and leaving time for us to do what we like and love?

There is a high risk that eventually several of humans would resign from their intellect and sit on sidelines, waiting for their smartphone app to tell them what to do next and how they might be feeling now!

When we invented automatic cars, an upgrade from manual gear shift cars, we almost lost hand-leg-brain coordination that was once important. It is not just a matter of coordination per se; it is the matter of **thinking**, which is what humans do and are good at, aren't we?

The enormous power carried by the gray matter in our heads may become blunt and eventually useless, if we never exercise it, turning it into just some slush. The old saying "use it or lose it" is explicitly applicable in this case.

Being present and engaging with the world around us can be a lost art if we don't pay attention to it. Million years ago, humans lost the tail, would the brain be the next?

Nothing pains some people more than having to think.

—Dr. Martin Luther King, Jr.[6]

Man plans, and technology laughs

It would be difficult or somewhat beyond our limits to predict what something smarter than us would do. A powerful system can be very difficult to control and stop. Many such concerns have been documented by Oxford Professor Nick Bostrom in *Superintelligence*[7] and by AI pioneer Stuart Russell.

The point I am driving here is not about evil AI per se; it is somewhat a red herring for now. But an incompetent machine that has not been developed responsibly and then has gone out of control is very much a reality even today. Without appropriate controls and applying sensible methods to this madness, the situation would only worsen further.

Misaligned or confused and conflated goals of an AI are going to be a significant concern of the future, as the AI will be extremely good at achieving its goals, but those goals may not represent what we wanted.

You may feel indifferent about birds and wildlife. However, if you are in charge of a project that is building a massive factory that could open up thousands of jobs, and in the process, you need to clear out a significant portion of a wooded area occupied by the birds—too bad for the birds.

The critical goal of responsible and safe use of AI is to never place humanity in the position of those birds.

▨ There is an old Yiddish adage, "Mann Tracht, Un Gott Lacht," meaning, "Man Plans, and God Laughs." I think a modern version of this adage might be "Man Plans, and Technology Laughs"! And we must avoid that from happening.

---

[6] *A Gift of Love* (Beacon Press, 1963).
[7] Oxford University Press, 2014.

Prevention

Avoiding AI risks proactively

Evaluating Risks of the AI Solution

Artificial intelligence, being such a powerful technology, inherently carries a high risk. It is a source of significant new threats that we must manage. This risk is not concentrated in one specific section of AI; instead, it is spread over all the stages and components of AI system. It includes AI solution design methodology, the AI solution itself, its development and deployment, and then ongoing use. Additionally, training of AI system and feedback loop also carry a significant risk.

Preventing each of risk means we will have to evaluate each component in its own right and seek avenues of prevention. Even before we discuss prevention, identification will be necessary. Figure 4-1 shows the six broad stages of AI solutions' life cycle.

To begin with, in this chapter, we will discuss risks related to AI solution. We will evaluate it from two angles—first to check if it is the *right solution* and second to verify if the solution is *designed right*.

If one of them is missing, that is, right solution but wrongly designed or rightly designed solution but fundamentally wrong—each will pose higher risks.

© Anand Tamboli 2019
A. Tamboli, *Keeping Your AI Under Control*,
https://doi.org/10.1007/978-1-4842-5467-7_4

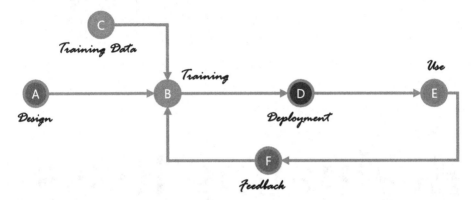

Figure 4-1. The six broad stages of AI solutions' life cycle

It starts with the correct hypothesis

When you want to verify whether you are working on the right solution or not, *hypothesis testing* becomes utmost important.

Typically, a hypothesis can be defined as an explanation for something. It can also be an assumption or a provisional idea, an educated guess that needs further validation.

Unlike our belief, which doesn't need validation or evaluation, a good hypothesis is always verifiable or testable. It can be proven to be true or false.

From a scientific perspective, a hypothesis must be falsifiable, which means there would be certain conditions or scenarios where this hypothesis or an assumption will not be valid.

Most importantly, you must frame the hypothesis as a first step before testing it or thinking about a solution. If you don't follow this rule, you might end up retrofitting a solution, which may not yield full results.

■ You must frame the hypothesis as a first step before testing it or thinking about a solution.

If you start with a flawed hypothesis, you would most likely end up selecting wrong solution, and several risks would seep in. Some of the risks associated with an incorrect assumption are financial loss, usually in the form of revenue leakage or operational losses, reduced customer loyalty and increase in frustration, risk to brand reputation, and several others.

You may be wondering, "Why is it so important to form a correct hypothesis?". Let us take an example.

A curious case of Acme Solutions

Acme Solutions (a hypothetical company) is experiencing significant growth in their products and services. Acme's customer base is continuously increasing. Increase in customer base is leading to increased number of calls to their incident management support center.

To keep up with the growth, Acme Solutions wants to increase head count in their incident management support center. However, acknowledging human limitations, Acme is keen on implementing artificially intelligent chatbot solution. By doing this, they expect to provide adequate support to incidents reported by customers, without additional head count. Customers will be able to interact with chatbot and report their problems; perhaps, get instant solution too.

Should Acme Solutions embark on AI chatbot project?

If this project is completed, by all standards, will it solve Acme's problem(s)?

If you think deeply, you would realize a flaw in Acme's thought process, their hypothesis. Their hypothesis is, if we implement AI chatbot, we can address increased number of the customer reported incidents, without increased head count.

What is wrong here?

The very first problem here is *"retrofit thinking."* Acme Solutions is starting with a solution in mind and then trying to fit in with their current issue of growing customer base and incident reports. They formed their hypothesis while they already chose the solution, which is the fundamental mistake in hypothesis framing.

Instead, if they focus on the problem first, the solution may be different. If Acme acknowledges that if an increase in customer base is leading to increased incident reports, then there is fundamental flaw in their service or product. There is something inherent in their service or product which is causing incidents in first place, and the best way to address it is to eliminate the cause of incidents.

Once Acme realizes that solving flaw in product/service would eliminate or reduce the number of reported incidents, their need for increased head count is likely to go away. When that happens, the need for AI chatbot would go away too.

In this case, if Acme were to go ahead with AI chatbot, their incident count will not reduce. However, they will be able to respond to more customers that are reporting incidents. Now, no matter how intelligent and efficient this chatbot is, once customer is reporting an incident, they are already unhappy and frustrated with Acme's product and service. Acme's brand loyalty has already taken a hit and solving customer problem wouldn't restore it; at best it will stop it from getting worse.

You will notice that AI risks do not necessarily start with the AI system or application itself. They would have already crept in, well before the solution building starts or deployment begins.

Therefore, correctly framing the hypothesis, validating it with the data, and supporting it with statistical analysis are highly essential. It should be your very first checkpoint in evaluating the risks of the AI solution.

Correctly framing the hypothesis, validating it with the data, and supporting it with statistical analysis are highly essential.

Does the solution address a valid root cause?

Once you frame the correct hypothesis and validate that you indeed need an AI system to fix something for you, it is time to decide, exactly what needs fixing?

Typically, we can establish a relationship between the problem and its cause in the following manner:

$$y = f(x)$$

where **y** is the problem and **x** is the cause or driver (or multiple causes or drivers) of that problem and **f** represents the process or formula that makes **x** result in **y**.

If we continue with our earlier Acme Solutions' example, it may look something like this:

> Incident reports = **f** (*poor product manual, low-quality product, etc.*)

It means incident reports are caused by poor product manual, which customers cannot understand and therefore create an issue. It also means that sometimes product manual may not be the only issue, poor product build may also cause an incident report.

So, if your solution were to address the *increasing incident report* problem, it cannot do anything to incident reports directly. It will have to work on the product manual, product build, and other contributing factors. When those factors are improved, automatically incident reports would reduce.

What if by implementing AI-based transcriber the product manual quality improves, which could write product manual automatically and validate it for content quality?

It would be a useful AI solution that is set to address a valid root cause. Your AI solution must be able to address one or more of the root causes of your main problem. It ensures that your issue will be solved.

If AI solution is not addressing any of the root causes, or if it is addressing only low impact root causes, leaving out highly impactful ones, that's a risk, and you must fix it. Make sure your solution is right.

Is the solution correctly trained?

One of the critical stages in AI solution development is the training of the system.

If you are working with a sophisticated solution, it will usually follow two-stage learning. In the case of a simple AI solution, it happens in only one stage, that is, during the model building phase.

The training stage is where foundations are formed for the first time and the solution is trained using those models. Therefore, the accuracy of these models is highly dependent on the training dataset.

If the data fed during the training of the solution is not good enough, it can result in several problems later on, and the risk would manifest in different forms.

For example, algorithmic bias may seep in the solution, which will affect all the outcomes. Machine learning algorithms identify patterns in the data and then build a predictive model based on that pattern. Any rule created or decision taken by the AI solution is dependent on that. If those patterns have any bias in them, then the solution will amplify that bias and may produce outcomes which in turn may reinforce the bias further.

The risk with such invisible bias is that when the AI solution makes any errors, it is tough to diagnose and pinpoint the source of an issue.

If you suspect the training quality of your AI solution or have doubts over the data used for training, the solution has higher risk of problems. On the other hand, if you are confident about the source data and training process followed, the risk is quite low.

There is no one and the absolute correct way to train the AI system; variations exist. If the confidence level in those training procedures and training data is low, then the solution carries high risk. Alternatively, if you are comfortable, it may be of low risk.

Has the solution considered all scenarios?

Following the premise of the solution that is *designed right*—one of the critical factors in evaluating the risks of the AI solution is the depth and breadth of design. You want to understand and be confident that the solution at hand is fully equipped to handle the entire spectrum of scenarios that may occur in your business.

The AI systems do not perform their tasks consciously. They rely on the training data, which means system's reliability and reliability of its outcomes are at risk if the input data had problems. These data problems are not just bias and quality of the content, but they are also related to the representation of various scenarios in the data.

Typical IT solutions, including AI and alike, are mostly point-based, that is, they focus on one goal to achieve with handful other parameters to optimize when achieving that goal. Unfortunately, they never can be all-inclusive and holistic; it is too much to expect from them.

Let us take the example of an autonomous car. If you ask the car to take you to your destination as fast as possible, it may get you there. Moreover, when you get there, you may have broken several traffic rules and have put fellow traveler's life at risk. None of that is acceptable. However, that may happen if your AI is too narrow and cannot see the wood for trees.

Many data with only a few types of scenarios cannot train the solution for real-life use cases, and this is a significant risk. Unless your training dataset represents all the possible use cases and scenarios, it is useless.

Since machine learning is an inductive process, the base model used by your solution can only cover what it has been trained on and seen in the data. If your training has missed low occurrence scenarios (also known as long tail or edge cases), your solution will not be able to support them. Your solution will be too risky in those scenarios or may fail.

If the data is missing variety, then it can lead to problems in the future. The solution developed with limited dataset and narrow training would be highly risky. Also, if the solution design has covered the breadth and depth of data and as many scenarios as reasonably possible, risks are lower.

Is there an option to stop gracefully?

A system that is *designed right* will always have redundancy inbuilt. New age AI solutions can be highly potent and may result in scalable damage when something goes wrong.

So, ask your development team, or your AI vendors, about the kill switch. Check with them the possibility to stop the system gracefully when needed.

A system without an option to stop gracefully is a highly risky system.

It may be the case that you have an option to stop the system when needed. However, exercising that option is not as simple as pulling the plug. That means you may still face a significant risk of jeopardizing business continuity in the process if you decide to stop it.

Ability to stop is not as bad as not being able to stop at all though. In this case, you must evaluate your options and alternate plan. What are financial and overall business repercussions of those alternative options when invoked? If stopping AI system in an emergency would mean you are required to put up 100-member team in place to cope with the situation, that could be a significant financial and HR burden for you.

When evaluating your AI solution on this parameter, check the availability of kill switch or graceful stop and restart options. If there are no such options, it is a high-risk solution. However, if these options are available with some level of inconvenience and minor losses, it may be termed as medium-risk solution. Although highly unlikely, there may be a solution that is easy to stop and restart. If it won't cost you too many resources while doing that, it may be low-risk option and should be the most preferred one, from risk point of view.

If you cannot stop an AI solution easily and gracefully, it is high-risk solution. On the contrary, if you are in full control and can exercise that option easily and without any significant loss, it is low risk, that is, better solution.

Is the solution explainable and auditable?

A software solution, no matter how sophisticated, is always auditable and understandable (explainable), unless it wasn't *designed right*.

Specific to the AI solution, explainability has been one of the most talked-about topics. With so many different approaches to develop AI solutions, it is getting increasingly difficult to determine how systems are making decisions. With explainable and auditable AI solutions, we should be able to fix accountability as well as make corrections and take corrective actions.

For many industries, AI explainability is a regulatory requirement. For example, with the General Data Protection Regulation (GDPR) laws in effect, companies will be required to provide consumers with an explanation for AI-based decisions.[1]

There are growing risks of AI when implemented without adequate explainability. These risks affect ethics, business performance, regulation, and our ability to learn iteratively from AI implementations. The dangers of unexplained AI aka. The black box is real.

People don't want to accept arbitrary decisions made by *some entity* that they don't understand. To be acceptable and trustworthy, you would want to know why the system is making a specific decision.

In addition to explainability, being auditable is another key for the right AI solution. Whenever something goes wrong, you would need to recover an audit trail of the entire transaction. Without being auditable, AI solutions are a significant risk. In the absence of verifiable records, regulatory fines may ramp up, and financial losses could pile up quickly. Not only that, it also becomes a limiting factor for you as a business. Without proper explanation and understanding, you cannot improve. Controlled and measurable improvements become near impossible when you are in the dark.

If the solution is explainable and auditable, the risk of wrong decisions and further ramifications is lower, and it is better for improvements in the business.

Is the solution tailored for your use case?

Most of the AI solutions claim to be personalized for the end users. However, the critical question is, well before you deploy or start implementing those solutions—are they customized or tailored for you?

For many use cases, geography-specific rules apply. For example, if you are a European business, your obligations related to data handling and explainability, verifiability, and so on are different as compared to other countries. If your business is in China, then facial recognition technology may be acceptable; however, for most of the other countries, it would be quite the opposite.

---

[1]www.fico.com/blogs/gdpr-and-other-regulations-demand-explainable-ai

If you expect your solution to transcribe doctor's notes and make prognosis or decisions based on that, your solution must understand local vernacular, slang, and other language nuances. For example, trash can (US) vs. garbage bin (AU), or sidewalk (US) vs. footpath (AU), mean the same things but have different words. The examples given here are quite rudimentary; however, for a sophisticated application, language differences would mean a lot.

In the case of a hiring solution or applicant tracking system (ATS), wouldn't it be relevant if the solution was trained on localized or geography-specific data during the initial stage? The hiring pattern and criteria are significantly different for different countries, which means the use of a standard solution is not a good fit for the purpose.

Geographical differences also influence user interactions with any software systems and so is the user experience. It can be best handled during solution design phase if the development team has local experts on board. These experts can provide valuable insights and knowledge to make the solution more effective and adaptable to the use case by end users.

Not having a tailored solution for your use case is not a deal-breaker in most of the cases. However, it is safer to assume that the relevance is essential to think about it and evaluate it.

If the solution is not tailored to your use case, you may be dealing with high latent risk. Evaluating the level of customization can help in estimating potential risks.

Is the solution equipped to handle drift?

The problem of change in data over time and thereby affecting statically programmed (or assumed) relationships is a common occurrence for several other real-life scenarios of machine learning. The technical term, in the field of machine learning, for this scenario is "concept drift."[2]

Here the term *concept* refers to the unknown or hidden relationship between inputs and outs. The change in data can occur due to various real-life scenarios, which can result in degraded performance of your AI solution.

---

[2]https://medium.com/tomorrow-plus-plus/handling-concept-drift-at-the-edge-of-sensor-network-37c2e9e9e508

If the solution is *designed right*, it would have accounted for such drift in the first place and therefore would be able to adapt to the changing scenarios. However, if that isn't the case, either your solution would fail, or the performance will degrade, which may also mean lower accuracy, lower speed, high error margins, and other such issues.

However, remember that it is difficult to identify any scenario in which concept drift may occur, especially when you're training the model for the first time. So, instead of working on pre-deployment detection, your solution should safely assume that the drift will occur and accordingly have a provision to handle it.

Your solution is not capable of handling drifts; you will have to fix it eventually on an ongoing basis. It is a costly proposition as you keep spending money for constant fixes. The risk of tolerating poor performance for some time exists. This risk increases if detection of drift is not apparent. If you, for some reason, could not detect the drift for a more extended period, it can be quite risky on several fronts. Thus, detection of drift is a critical factor too.

Check if your solution has accounted for concept drift and if yes, then to what extent. Answer to that check would tell you the level of risk.

AI solution evaluation questionnaire

Here is the questionnaire, which you can use in evaluating risks of your AI solution. It is applicable in all cases, whether you are developing the solution in-house, have outsourced the development, or are buying it off the shelf from another vendor:

1. Did you frame the hypothesis correctly?

2. Did you validate the hypothesis with data, and does statistical analysis support it?

3. Collectively, how many root causes does your AI solution address?

4. Does your solution address the high impact root cause(s)? Ideally it should address all the drivers, but if not, it should at least treat the most important ones.

5. Do you know how your solution was trained? How many times or how long and when?

6. Do you know what type of dataset was used during the training of AI system? Wherefrom the data was sourced? How reliable do you think the data and its source are?

7. How was the fitness for purpose assessed for training dataset?

8. Does the training dataset fully represent the population under consideration?

9. Does the training dataset contain biases? If yes, of what kind? If not, how was that ascertained?

10. During the initial stage, did you perform dataset cleaning and normalization? Did you find any anomalies? Are you aware of the source of those anomalies? How were those corrected before training?

11. Does the training dataset contain a full spectrum of values? Are any long-tail values discarded? If yes, how did team come to that decision?

12. Can AI solution be stopped quickly and gracefully as and when needed? Do you know the process to do it?

13. Once stopped, can the solution be redeployed? What is the process to do it?

14. Can your AI solution forget all the learning and relearn from scratch? What resources are required to retrain the system? If an AI solution cannot unlearn and retrained, how can you fix any issues?

15. When the solution is stopped, how would business function? Do you have a business continuity plan in place? Have you evaluated financial repercussions, downtime, claims, and other impacts relating to this?

16. Is your AI solution tailored to your use case? To what extent? If not fully customized, what kind of crossover is expected? Do you have plan to handle that?

17. Has the solution been tested for concept drift? If not, why not? If yes, how?

18. What precautions are built in the solution to handle drift? Do you know how to identify drift? Also, how to fix it?

19. Are you able to explain and understand your AI solution's decisions and predictions?

20.   Are you able to perform a full audit of your AI solution?

21.   If something goes wrong, how long will it take to audit and trace the root cause of failure? How does it impact the business?

Taking a cue from these questions, you may think of more questions to seek answers. In the end, it is right for you because this way you can be more confident with your AI solution and understand it better.

Evaluate thoroughly and be in control

From an economic point of view, currently there is significant information asymmetry that has put AI solution creators in more powerful position than the ones who would use it.

It is being said that technology buyers are at a disadvantage when they know much less about the technology than sellers do. For the most part, decision-makers were not equipped to evaluate these systems. However, with the preceding questionnaire and information, the situation should change.

The crux of the matter is, you cannot leave it to the AI developers, in-house or outside, to police themselves for mindful design and development.

It is natural for us to evolve and keep pushing the envelope. However, certain things cannot be undone. There is no *undo* option here. Therefore, it is imperative that we either slow down or be overly cautious.

Since slowing down is a less likely acceptable option, being overly cautious is necessary. Being strict in evaluation of risks and putting up strong governance are some of the critical steps you can take.

It may be a minor nuisance if your laptop crashes, or your search engine shows weird and unrelated results, or your movie streaming service suggests you flop movies. However, if the system is controlling a car, or a power grid or a financial trading system, it better be right. Minor glitches are unacceptable.

For you to be confident in your AI system, you must make sure that you are working on the *right solution,* and the solution is *designed right.*

Tactically, "doing things right" is essential; while strategically, "doing the right things" is critical. With powerful technology like AI at hand, both are essential.

De-risking AI Solution Deployment

In the previous chapter, we focused on AI solution design and risk aspects related to it. Our objective was to ensure that the solution is *designed right*. In this chapter we will focus on solution's deployment aspect and also cover *solution is right* aspect.

When you have access to powerful tools and technologies such as AI, deploying the right solution in for the right problem at the right time becomes almost mandatory. It hasn't been any different in the past. However, due to the scale at which the AI can now make an impact, it demands a higher level of sanity and responsibility.

The fact that *you don't know what you don't know* poses a significant barrier when dealing with AI like solutions. The best way to break that barrier is asking the right questions and seeking answers.

© Anand Tamboli 2019
A. Tamboli, *Keeping Your AI Under Control,*
https://doi.org/10.1007/978-1-4842-5467-7_5

For AI solution deployment, it is essential that you take control of planning and architecture early on. Avoiding asking right questions early on can mean you end up with subpar solutions with reduced yield. Very likely, you may waste your investment altogether by doing something that doesn't affect your key metrics.

If you ask for tough questions early on, it may seem like you are delaying; however, remember that it is for the greater good. It is an oft case with new products and technologies such as AI, novelty sells! You do not want to buy a fancy and costly toy system. You want to set the foundation to boost your business significantly.

Ensuring long-term strategy is defined

I often say that technology solutions are not one-point solutions anymore; they are intricate pieces of systems. Therefore, expecting your AI solution to be plug-and-play or install and get-go would be wrong.[1]

You must ask yourself and your leadership team how your strategic road map looks like or should look like after deploying AI solution. If you do not have any such road map, that's your first and foremost priority, work on it already. Without having a strategic road map in place, there is no point in deploying any solution whatsoever.

The use of emerging technologies such as AI is a direction and not a destination. Therefore, your road map should show the evolution of your business as you imbibe these techs. Having this information mapped out on a stage by stage manner would be highly useful.

For each milestone and critical step, make sure you document which business metric(s) will improve and what is the expected outcome.

This strategic road map can then be your ultimate guide—a north star for your entire journey of technology deployments. This type of document will naturally evolve as you make amendments in your strategy and expectations from time to time.

Defining your problems correctly

In the previous chapter, we covered this point partly under *importance of hypothesis*. A hypothesis is explanation of something or assumption that needs further validation.

---

[1] https://medium.com/tomorrow-plus-plus/5-key-questions-to-ask-before-your-emerging-technology-journey-begins-9bcc025417e4

However, if you do not correctly define the problem in the first place, any assumption or explanation formed after that would not help. Correct hypothesis on wrong business problem or for the one of the least priority is not going to yield significant benefits.

If you have a right problem to be solved, but you make a mistake in framing the hypothesis, it may lead to a waste of resources including other side effects on your business; the keyword here is "may." However, if you choose a wrong (or less critical) problem to solve, correctly framed hypothesis won't help; you are already on a wrong path here. Therefore, you will waste resources and have negative impact on business; the keyword here is "definitely."

The hypothesis is like a vehicle, while problem is like a destination. If you have chosen a correct destination (i.e., problem to solve), poor choice of vehicle will take you to your desired destination a bit late, or the right choice will take you there faster. Either way, you will reach your destination, that is, problem will be solved. However, if your destination is incorrect, no matter how good the vehicle is, it will be pointless.

Defining your business problems is mostly a strategic exercise, you can evaluate what your goals are and what are the things stopping you from achieving those goals. You may treat those impediments as problems to be solved. Once you've listed your problems, quantified, and validated them properly, you can frame the hypothesis for each.

Remember to define the problem(s) correctly as a foundational requirement before jumping to the hypothesis framing stage.

Problem is your destination, whereas hypothesis is your choice of a vehicle.

Validating all the root causes

If your AI solution is addressing your validated root causes, it will most likely be the right solution. However, important question would be—how do you know your root causes are valid?

I always recommend a two-way process for root-cause validation. This process ensures that you have chosen valid root causes and can make a significant impact on final outcome.

The first part of this involves traditional statistical and analytical methods. These methods mainly include statistical hypothesis testing and five-why analysis. Applying these techniques would lead you to short-list potential root causes; some of them would be significant, while some may not. It would be best if you treated these root causes as mere drivers of the main problem.

Not all causes are sole drivers of the problems, and thus eliminating them may not necessarily result in problem-solving. That's where the second part of the process comes in handy.

By applying reverse inference

The second part involves *reverse inference* technique. This technique helps you to demonstrate the logical fallacy (if any) of affirming what you found as a problem driver(s). Doing this will either prove that a specific outcome (problem) is generated solely by a particular occurrence (of driver) or it will disprove it. If disproved, you can be sure that removal or reduction of that root cause is not going to solve your problem. On the contrary, if the reverse inference proves that a specific occurrence of given root cause is the only definitive path for problem to occur, then you can be assured that solving it will solve the main problem.

For example, a poor customer rating is a result of the contact center interaction full of friction, poor quality of product manual, and lousy product quality itself. If you apply reverse inference, making contact center interaction frictionless *may not* necessarily improve ratings. However, improving product quality can *certainly* lead to better ratings. Thus, the primary root cause here would be the product quality which you must enhance for significant impact on final outcome (ratings).

By applying the MECE principle

Root causes also need to be following *mutually exclusive collective exhaustive*, aka MECE principle. That means while the optimum arrangement of all the problem drivers is exhaustive and you've included all drivers, you do not double count drivers of the problem which overlap or have same effects. Ensuring MECE means you are optimizing your resources in solving those issues and there is no overlap in their work. It will help you avoid deploying two solutions, which address the same problem and thus making one of them redundant.

Validate all the root causes found by using the two-step process, and you can be confident that your *AI solution is right*.

Planning for business continuity

Being skeptical is the key to de-risking your AI solution deployment.

Think about business continuity planning (BCP), just in case these solutions do not work out as expected. I have seen several projects that do not get delivered on time or do not get delivered at all, which calls for operational rollback to the status quo.

This planning is required not only during the deployment but also on an ongo-ing basis. If for whatever reason you were to take the AI solution offline, can your business function as usual? If the answer is no, make sure you have an alternative in place before deployment starts.

Clarifying the accountability

From the project management perspective, fixing an accountability is rela-tively easy as there is method to it. However, when something goes wrong during AI solution deployments, it is not a black-and-white situation. While the solution may be behaving abnormally (indicating vendor's fault), it may very well be doing that due to incorrect training data input to it (indicating your responsibility). Moreover, since things can't be so precise in AI deployments, clarifying and fixing accountability is essential.

Your team must draw out all the potential failure scenarios during the deploy-ment and assign accountabilities to respective parties and get their sign-off on it. Doing this can ensure that if and when implementation faces hiccups, they are fixed without any blame game and waste of additional resources.

Ensuring the availability of the right infrastructure

It relates to your preparedness from a technological standpoint. While your business leadership might be ready to accept the challenge and get started with the AI implementation, your current infrastructure may become one of the weakest links of the critical chain.

Get a realistic view of the state of your current IT infrastructure and capabil-ity. Augmenting IT infrastructure for AI projects is not a *click and buy* affair and requires significant resources to get ready, hardware and networking included. If there is a gap or you are not confident about its fitness for the purpose, fixing that first will be a prudent step.

Process maturity is another aspect that you must focus on. In addition to the IT infrastructure, having your operational processes at a significantly mature level is highly recommended. If that is not the case, there would be several iterations along the way as your requirements would continuously change and can lead to many improvisations on the go. These iterations and changes clearly will put a dent on the budget and can delay the implementation beyond acceptable limits. The change fatigue involved as you keep doing this could soon drain energy from people and demoralize them. It can quickly become a significant issue.

Data readiness is the third critical aspect of de-risking AI deployment. A simple way to look at it is this—do not think about AI solutions of any size if you have questions about your data capabilities. AI solutions demand a massive amount of data to perform at acceptable levels. However, if you do not know what to collect this data, then it can be a significant problem. Make sure that you are data ready in all aspects, well before thinking about AI solution implementation.

Setting up target metrics of success

Like every other project, AI project will also have a honeymoon period for a while. Once it is over, your stakeholders will be interested in real, measurable business value.

If you have not identified the key metrics up front and linked to the outcomes of the project, this could be an issue. You should identify and validate the impact beforehand, way before you start the project so that you will know what to expect and when. It will also enable you to pull the plug and stop the deployment if it is not adding any value to your crucial metric or not going as planned.

Implementing a "cool" technology in your business may give you some bragging rights, but it wouldn't contribute to your top line or bottom line in its right. There should be a clear linkage of each business metric with each planned initiative. You must establish this linkage as a first step when you start the AI project. Establish how you will measure the change in metric, what will be an acceptable change, and what will be your lower limit.

Having this measurement system will help you march forward or to take a pause and rework as you progress through.

Defining your acceptance criteria

Before deploying your AI solution, you must set your acceptance criteria for the deployment outcomes. Doing this beforehand helps in avoiding any confirmation bias and also can save you from facing endowment effect.

Endowment effect states that people are more likely to retain an object they own (say AI solution) than acquire that same object when they do not own it.

So, once you deploy the AI solution, you are more likely to relax your acceptance criteria and accept a subpar solution, simply because you've now invested resources in implementing it. Once this happens, you will have to live with dysfunctional or a poor-quality AI solution, forever. You may argue that there would be ongoing improvements to it, but the fact is, you would always find yourself catching up, and it is not a good situation to be in.

▒   If you accept a subpar solution with the hope to improve it later as you proceed, you will end up playing catchup with technology, and it is not good.

Defining acceptance criteria to fit the output is nothing but retrofitting and can put you in massive intellectual debt. Intellectual debt is a situation, where you accept the answer, but don't know how it was derived and expect it to be known later point in time. Getting into intellectual debt can increase your risk exposure massively and put you in weakest of spots.

If you want to de-risk your AI deployment, define your acceptance criteria first and stick to it. If results are not stellar or do not meet minimum passing/ acceptance criteria, reject the solution and do over. It is better than exposing your business to greater (unknown) risks.

Building the capability for execution

Deploying AI solutions demand a wide range of skills and in-depth knowledge of several aspects of the business to see successful outcomes. If you do not have the right capability within the in-house team, it could become an impediment. Nonetheless, if you identify this gap early on, you can work toward bridging this gap.

Generally speaking, it is ideal to have talent and capability available in-house as a part of your core business team. However, often, it may not be the case, and if that so, getting external support is the next thing you should do. Here it would be best if you lay your expectations when looking for external help. It is necessary to source the right support.

There is one thing you should watch out. Your AI vendor(s) should only be responsible for the timely and flawless delivery within specified constraints. The product and project ownership, as well as accountability, must remain with your internal leadership team. If you compromise on this aspect, it may lead to adverse consequences. So if this capability (to lead) is lacking, fix it before moving further.

Handling security and compliance aspects

AI solutions are usually black-boxed and are known for nasty surprises. Their detrimental impact could cost you a fortune. They may not only have significant operational negative implications but may also cause considerable reputational stress. Depending upon the size of your brand and company, the level of risk may be higher or lower.

Note that cybersecurity is slightly different from AI security, and it must be acknowledged as you plan for AI solution deployment. Having the right expertise at disposal from the beginning will ensure that your implementations do not have any loopholes and are defensible from various possible threats.

Having audit functionality in your AI solution is essential to have. It is not only necessary from operational perspective but also required throughout the deployment journey. Which means, at any point in time, when deploying solutions as well as after you have implemented them, you should be able to see what happened, good or bad, how and why. Having a fully compliant, auditable, and retraceable system means your business is covered in case of negative consequences.

Integrating with your other systems

AI systems do not work and stay in isolation, and even simple interactions can lead to big trouble.

Here is a classic example of two racing algorithms that ended up listing a book for upward of $2.1 million on Amazon in 2011. Michael Eisen, a biologist, one day found from his students that an unremarkably used book—*The Making of a Fly: The Genetics of Animal Design*—was being offered for sale on Amazon by the lowest-priced seller for just over $1.7 million, plus $3.99 shipping. The next cheapest copy weighed in at $2.1 million. The respective sellers were well established; each had thousands of positive reviews. When Eisen visited the page the next day, the prices had gone up yet further.[2]

Eventually, they realized how two sellers were trying to compete with each other systematically and therefore landed in this situation.

In another example, thoughtless automation deployment resulted in increased number of customer complaints. When one of the tier 1 telecommunications company deployed automated system, the system was automated to close the trouble tickets within the specified period. Tier 1 company expected that tier 2 company would reopen the trouble ticket if there were an issue faced by the customer and it will continue using the same trouble ticket until the resolution. However, the tier 2 company also decided to automate their systems (thoughtlessly), where the system would close the trouble tickets soon after the tier 1 company closed them. Doing this meant—if the issue persisted—customers were required to lodge new trouble ticket, and it would be counted as an additional trouble ticket, although it would be the same in reality.

---

[2] www.michaeleisen.org/blog/?p=358

Since both the systems were racing to close the ticket as soon as possible, they missed the validation part and caused an increased number of customer complaints. This issue hurt their reputation badly and stressed operational metrics beyond repair. Both companies had to lose (or waste) money and other resources to deal with this condition.

These were two simplistic examples of what can happen when multiple systems interact with each other, and there is rise of collective stupidity out of this.

When you plan to deploy your AI solution, you must evaluate all the integration points and ensure that they are stable. Making sure that you have proper testing and validation plan for those integrations can save you from unknown risks and losses.

Several times you may have to make make-shift arrangements in joining two systems for interaction and making sure that they work cohesively. These are the weak spots that you must monitor during and after the deployment.

AI solution deployment questionnaire

Here is the questionnaire, which you can use do de-risk your AI solution deployment. This questionnaire will help you answer crucial questions and highlight any gaps or risks related to AI deployment:

1. Do you have a strategic road map for implementing AI systems in your business? Does it (strategic road map) have clearly defined critical steps and milestones?

2. Did you define your core problem correctly before thinking of AI solution? How have you ensured that it is the correct and essential problem?

3. How does solving this problem fit in your strategy? What are the benefits you stand to achieve by solving this problem?

4. Did you validate all the root causes driving the given problem? Did you confirm these root causes with reverse inference technique?

5. Does your list of root causes follow MECE principle?

6. Do you have a business continuity plan ready? Have you clearly defined accountabilities and responsibilities relating to BCP? Does every accountable and responsible person know about it? Do they know what do you expect from them?

7. Do you have the right IT infrastructure in place for implementing AI systems? Is your IT hardware and network capable of running sophisticated AI algorithms on-premise? How would you proceed if there are issues with IT infrastructure?

8. Is the process mature enough, where you are deploying the AI solution? Are you anticipating any significant changes to this process?

9. Do you have a system in place to generate or gather required data for the AI solution? Do you have adequate and proper data storage facility? Do you have a data management plan? How would you proceed if there are issues with the data availability?

10. Would the delay in project timeline impact dataset validity (freshness) for the AI training? If yes, what are the impacts? If not, why and how doesn't it impact?

11. Have you set up the target success metrics? How do you plan to measure them? Has everyone agreed to these metrics and measurement methods?

12. Have you set up acceptance criteria for the AI rollout? What is your action plan if it is not acceptable? What is the minimum limit for being acceptable? What is the upper ceiling to accept the solution? What if the AI solution is merely passing the criteria?

13. Are you aware of internal capability requirements for AI deployment?

14. Do you have an internal capability to see through the AI deployment? How do you plan to bridge the gap wherever it exists? Do you have budget to source external support?

15. Would it be possible to audit the system deployment at each step? If yes, how and if not, then why not?

16. How are you ensuring legal and policy compliances of the system?

17. How are you ensuring the security of the system? Is there a way to modify security restrictions if needed? If yes, how to do that?

18. Have you accounted for all the interaction points of your AI solution with other systems? How many make-shift arrangements are needed to make things work seamlessly? How do you plan to mitigate risks at those interaction points?

19. Are there controls and alarms in place, if the system interaction digresses from the routine? If yes, what are thresholds? Can these thresholds be updated?

Asking the right questions is a must

If you are set to ask these questions before beginning your AI deployment, there is much broader debate each question, and its answer will trigger. How you drive it to its logical conclusion is all up to you. If you do that in the right manner, it will give you enough confidence to move further.

However, if you do not get satisfying answers, it will prompt you to go back to the drawing board and redo the strategy and planning for good.

AI solution deployments are more complex than typical IT software deployments and, if not treated diligently, can leave a bad taste for the rest of their operational life and then some.

Well begun is half done, and asking the right questions is part of it!

Good AI in the Hands of Bad Users

Until now, we have discussed generic and specific risks of AI solutions that were either applicable at a macro level or were specific to the solution design. In this chapter, we will talk about risks arising out of the interaction of the AI system with other systems and human users or operators.

Usually, the AI systems are not stand-alone; they often interact with several other systems and humans too. So, at each interaction point, there is either a chance of failure or degraded performance.

If the users or other interacting systems are not good enough, then no matter how intelligent your AI system is, it will eventually fail to deliver. The failure may not be the only outcome, but in some cases, it may have also resulted in the business risk.

▓ Your AI solution is only as good as its users.

© Anand Tamboli 2019
A. Tamboli, *Keeping Your AI Under Control*,
https://doi.org/10.1007/978-1-4842-5467-7_6

What does a bad user mean to us?

Typically, one can classify computer users by their roles or expertise levels. In the case of role-based classifications, they look like administrator, standard user, guest, and so on, whereas skill-based groupings put them in the following categories: dummy, general user, power user, wizard/geek or hacker, and alike.

All these categories have user levels that are just good enough to use the computer or any software installed on it. However, if the user's expertise is a borderline scenario for being good enough or below that they would soon become bad users of technology. So much so that they can cause a relatively good computer system to come to a halt, AI included.

Additionally, I have also seen the following user categories; each of these is dangerous enough to cause problems:

Creative users

Creative users are generally skilled enough to use the tool, but they often use it beyond its specified use. Doing that may often render the tool useless or break it.

I remember an interesting issue, during my tenure with LG Electronics. One of the products LG manufactured was a washing machine. A typical home appliance that normal users would use for washing clothes!

However, when there were several field failure reports from service centers, especially from the North-West part of India, we were stunned by the creativity of washing machine users.

Restaurant owners in Punjab and nearby regions in India used washing machines for churning the Lassi at large scale.

Lassi is a popular and traditional yogurt-based drink that originated in the Indian subcontinent. It is a blend of yogurt, water, spices, and sometimes fruit. Due to its thick texture, churning Lassi requires more human strength, especially when one is making it in large commercial quantities.

That's when restaurant owners became creative and used top loader washing machines for Lassi churning. It caused operational issues due to unintended and unspecified usage of the appliance and resulted in an influx of a large number of service calls. This kind of creativity looks interesting at face value but certainly causes problems with the technology tools.

Another example of such creativity would be the use of Microsoft Excel in organizations. How many companies have you seen where Excel is not only used for tabulation and record-keeping but also being used for small-scale automation by running macros?

How many times you've seen people using PowerPoint for report making instead of creating presentations?

All these are creative uses of tools and maybe okay once in a while. However, the users are mostly abusing the system and tools which can cause unintended damages and losses to the organization. These types of users also expose companies to more substantial risks.

Mischievous users

These types of users are not productive, and they do not mean any direct harm. However, they are merely toying with the system. At times, this may be completely harmless. However, they may cause unknown issues, especially with AI like systems. If your AI system has a feedback loop where it gathers data for continuous training and adjustments, this may be an issue as any erroneous or random data can disturb the set process and models.

Deliberate (bad) users

The users that are deliberately acting bad and trying to sabotage the system could be disgruntled employees.

Sometimes, these types of users think that AI system is no better than them and it must be taught a lesson, so they become deliberate in their attempts to fail the system at every chance they get.

Mostly deliberate bad actors do it with some agenda. These types of users are difficult to spot early on.

Luddites

A classic example of bad users would be *Luddites*. They are people who are, in principle, opposed to new technology or ways of working.

The Luddites were a secret oath-based organization of English textile workers in the 19th century, a radical faction which destroyed textile machinery as a form of protest. The group was protesting against the use of machinery in a "fraudulent and deceitful manner" to get around standard labor practices. Luddites feared that the time spent learning the skills of their craft would go to waste, as machines would replace their role in the industry.

Over time, however, the term has come to mean someone opposed to industrialization, automation, computerization, or new technologies in general.

These users mostly employees who are threatened and affected due to implementation of new AI systems. If your change management function is doing any good job, these types would be easy to spot.

Bad user vs. incompetent user

Incompetence can mean different things to different people. However, generally, it indicates the inability to do a specified job to a satisfactory standard.

If a user can use the system without any (human) errors and the way they were required to use it, you can call them competent users. Incompetent users often fail to use the system flawlessly on account of their ability (not system's problems). These users often need considerable help from others to use the system too.

Bad users, on the other hand, maybe excellent at using the system, but their intent is not a good one.

All incompetent users are inherently bad users of the system; however, bad users may or may not be incompetent. The reason why we need to understand this distinction is—one is curable, while other isn't. You can make incompetent user competent by training them, but no amount of training would refrain intentional bad users.

Handling with change management

Most of the user and other system interaction–related issues are the result of poor or no change management during the full term of the project.

While AI has the power to transform the organizations radically, substantial adoption numbers are difficult to achieve without having an effective change management strategy in place. All the bases should be covered before you begin the implementation and continue it as long as it is necessary.

When you have a complete understanding of how an AI solution will help end users at all levels in the company, it becomes easy to convey the benefits.

Only quoting feature list of your new AI solution will not help, you will need to explain what exactly AI solution is going to do and change and how it will help everyone in doing their job more effectively.

A safer and less risky approach will be to pick tech-savvy users for the first round of deployments. They cannot only provide useful feedback about the AI system you're deploying but also can highlight potential roadblocks for a full rollout. Tech-savvy users can help you determine if the AI solution works as expected for their purposes.

These users then become your advocates within the organization and also help in coaching their peers when needed. These early users help in creating significant scale buy-in within teams and potentially reduce the number of bad users down the track too.

Educating users for better adoption

Proper training plan and established training use real-life scenarios and hands-on sessions—user feedback is welcome and will be acted upon before moving forward—not doing that means you leave out much unhappy right talent.

If you want to ensure smooth transitions and user adoption, start user education early in the process. Moreover, tailor it to each stakeholder group. You should provide them with baseline information and knowledge around AI technology as a whole and then deeper insights and information on the specific application that you're deploying. It will help in setting their expectations. It is essential that every involved member understands the benefits of AI solution.

By educational initiatives, you can quickly dispel misconceptions about AI. For some of the stakeholders and users, especially the ones unfamiliar with how AI can help, futuristic technologies can be intimidating. This intimidation begets a defensive response and brings out the lousy user in them in various forms.

With proper education, the benefits of AI can become apparent to your team members and thus foster positive uptake.

If you center the education on the fact that AI solution will enhance employee's daily work and will make it easier to handle routine tasks, make sure you highlight that aspect. When communicating with your employees, focus on the purpose of the change and emphasize the positive outcomes it can bring.

Even for executive leaders, it is vital to understand what is happening and knowing the capabilities or limitations of the AI system you're deploying. By investing due time in acquiring appropriate education, executives will be able to ask the right questions at the right time and steps. Being more involved is necessary for them.

It is hard to recover from a lack of end-user adoption if you haven't invested enough in user education, so make sure you have spent adequate budget in educating users for better AI adoption. Create multiple formats that are readily available for various devices, including offline in-person sessions. When training is rolled out, measure the uptake and types of resources employees use most. It tells you which medium is more effective, and you can leverage it some more.

Going all out on education and training materials can minimize the chances of failure when employees start using the systems. It will ensure that all the promises efficiency and productivity of AI solution are fulfilled.

There is a typical spike in productivity loss when new systems are deployed, which is generally a result of slow adoption and long learning curve. You can minimize this productivity loss with a proper approach. An approach where education and planning are paired with training is the best way to ensure successful AI deployment.

Moreover, as a rule of thumb, education and training should not end after solution deployments but must become a periodic activity to ensure that you can sustain all the positive gains. It will also help in improving user capability over time and help in reducing bad-user phenomena.

Checking the performance and gaps

It is reasonable to expect a human user to demonstrate the same performance repeatedly for any given set of scenarios. You would also expect them to reproduce the same outcome each time those scenarios frequently occur. Moreover, you would also expect other connected systems to exhibit similar consistent behavior for you to be able to trust the whole system.

It is essential to check performance for consistency and find any gaps as early as possible in the deployment phase. AI systems usually work on proportional outcomes, and some variation at the solution level is already accepted. When you couple this inherent variation with the variation of several humans and other systems, it can quickly become unmanageable. Although each variation might have been acceptable at an independent level, altogether it can be problematic and result in poor overall performance.

That's the reason why performance must be checked for these gaps once you deploy the AI solution. When you use your AI solution, several systems interacting with your AI solution may go haywire. If you did not plan for systematic changes before the deployment, then it could soon become a roadblock.

Performing *Gauge R&R* (gauge repeatability and reproducibility) tests can reveal several actionable findings. It is a statistical test used to identify variance between multiple operators and can be used to test how various users interact with the same system. You can also use it for checking how multiple systems interact with your AI solution.

The outcome of Gauge R&R studies gives you an indication of the causes of variation in the performance. These findings can help in formulating training plans for fixing user performance. It can also help you in formulating system change requirements to make them work seamlessly.

Continuously monitoring the user and system interactions and periodically conducting systematic checks (and tests) can help you in managing incorrect usage of your AI solution.

Do not forget Gemba visits

If you want to learn user behavior in real time, learn from it, and improve your systems and processes, Gemba studies are very useful.

Gemba (also spelled less commonly as Genba) is a Japanese term which means *"the real place."* Japanese police refer to a crime scene as Gemba, and TV reporters often refer to themselves as reporting live from Gemba.[1]

In business, however, it refers to the place where value is created. We have commonly used this term in manufacturing, where we referred the shop floor as Gemba. In other contexts, it can also be known as the site for a construction scenario, the sales floor for retail, or service lounge in the case of banks' customer lobby.

From the *lean manufacturing* perspective, Gemba is a highly useful methodology to uncover problems in the process and usage of tools or systems. The best improvements can often come from going to the real place where precisely the stuff happens, where you can see for yourself, the state of the process. The core principle says—go, see, ask why, and show respect.

Doing this enables you to see things firsthand, remain curious as you ask why, and understand your users more deeply. Maintaining a respectful approach can help in gaining employees' trust and letting them know your intentions to understand and fix the system as needed.

Remember that this is not management by walking around or performing inspection or audits of any kind. The objective of Gemba visit is to understand improvement opportunities.

Here are a few steps to conduct a successful[2] and productive Gemba visit that can help you identify improvement opportunities and improve users' performance too:

---

[1] http://insights.btoes.com/resources/what-is-going-to-gemba-lean-kaizen-definition-introduction
[2] http://insights.btoes.com/business-transformation-operational-excellence/be-a-rebel-with-a-cause-7-steps-for-a-successful-gemba-walk

1. ***Start your visit with a hypothesis:*** Such as "I believe that our AI solution is helping employees in improving their productivity, let's go and see how", or "Our AI solution is easy to use and users love it, let me see it for myself." With such hypothesis, you will go with a frame of mind to seek some inputs, and it would reflect in your interactions with users too.

2. ***Formalize a list of questions:*** Have your questions handy when visiting. Your questions could be leading ones to validate (or invalidate) your hypothesis. You may also focus on usability issues, frequent errors, and near misses observed by users or system performance problems such as speed or UI-related matters. These types of issues do not get reported frequently, or sometimes there is significant lag; your Gemba visits can uncover any such issues faster.

3. ***Share your experiences and feedback:*** As you walk through, make sure that you are interacting with users, commenting and expressing your views with examples and constructively reinforcing your company's reason behind deploying AI solution.

4. ***Ensure follow-up actions are recorded:*** If can be quite helpful in tracking every follow-up action if logged in the format of "what, who, and when." It can help in creating accountability and fix issues if any. More importantly, it can help you in improving your user's learning curve and enable you to reap the benefits of your AI solution quicker.

5. ***Establish a rhythm:*** Making sure that you are regularly doing Gemba visits can help you set a rhythm. How often you do these visits depends on the severity of problems you are facing in AI deployment, your user education objectives, and your findings from prior Gemba visits.

Remember that the objective of Gemba visits is to improve systems and processes. By fostering learning from feedback mentality, you can do it much faster. These steps are critical to avoiding "Good AI in the Hands of Bad Users" situations.

Handling user testing and feedback

No matter how much content you put into training material, it is not always possible to cover all the questions users may have. It makes it essential to establish easy to use and quickly accessible communication channel between users and responding team.

If you can make it clear who the contact person is, how long it will take to get a response and how to escalate if needed, that would help in gaining user confidence and giving them clarity about AI deployments. By doing this, you will only encourage users to come to you when they encounter any issues.

Giving them confidence that their feedback is valuable and you will always take it on board can go a long way. Moreover, once received, do not just consume the feedback, but act on it.

Sincerely learning for every feedback and fine-tuning your AI application can help in improving user experience. It can give them confidence in the deployed AI system. Doing this also reduces the number of bad and incompetent users significantly and thereby reduces your overall risk exposure quickly.

Establishing HAIR department

Thus far, the HR department has been carrying out responsibilities to manage the performance of (human) workforce. All the policies that HR teams have developed were there to address human workforce education, augmentation, and performance management.

This set scenario is changing as machines are becoming smarter, and AI is becoming mainstream. So, how do you plan to handle this new type of workforce, which is fully automatic (AI only) or is augmented by smart machines (humans + AI)?

HR teams will have to manage performance gaps and issues related to system malfunctions as well as retraining requirements of humans and machines. If there is any impact on human performance due to poor quality AI systems, it will have to be handled differently than how they would handle typical human (only) performance improvement.

Generally speaking, AI systems are smart, but they seriously lack the pivotal characteristic of human, common sense! With the deployment of digital twins of your human employees, it may become an essential requirement.

Humans in charge of powerful technologies would have to be trained, coached, and managed effectively just the same as the government controls armed forces differently than civilians.

It would be a good idea to take steps toward establishing a new HAIR team or augment existing HR team and accommodate these new challenges. Development of appropriate policies and procedures would be core to their initial tasks.

User readiness questionnaire

The following questions will help you in charting out user interaction and development plans effectively:

1. Do you know the profile of your AI users? Which users do you think need education and competency support? Which users you believe could be challenging for successful AI deployment and why?

2. What is your plan to get all the users onboard?

3. Do you have a change management plan in place? What is it pivoted on?

4. How does your education and training plan look like?

5. How do you plan to check performance gaps among multiple users and systems? How would you ascertain exact reasons for the discrepancies? Have you communicated this plan everyone?

6. Did you plan for Gemba visits? Who are the participants? What are its objectives? Are they aware of it?

7. How are you going to collect and act on user feedback? Have you established a single point of contact for feedback? How long will it take to get a response? Do you have a set escalation process, if needed?

8. How are you going to maintain continuous communication with your users?

9. What is your strategy to augment the HR department and increase its scope with AI? Have you updated performance measures in light of AI twins for users? Did you establish evaluation criteria to identify and ascertain improvement areas? How would you decide whether the system needs improvement or the employee?

Nailing the technical element is not enough for AI

No matter how smart the technology or AI in particular is, it cannot apply common sense and human perspective.

Therefore, merely nailing the technical element of AI is not enough; you need to balance it with the human aspect. The understanding of the world and the surrounding environment in which you are using the AI is crucial.

Increasingly, technology teams need to demonstrate cognitive intelligence if they were to be successful. As much as the development and deployment of an AI solution are critical, the user aspect is important too. Without proper use (and users), AI success will hang by a thread.

A good AI solution in the hands of bad users can be disastrous, while an average AI solution in the hands of good users can be a great success. Technology users have the full power to make or break it; your goal should be to enable your tech users and extract maximum positive value out of it.

▨ The people who understand AI users don't understand AI design. The people who understand AI design don't understand AI users.

A Systematic Approach to Risk Mitigation

Risk mitigation is not an easy task; instead, it is even harder if there is no quantifiable measure attached to it. In this chapter, we will discuss a systematic approach to risk mitigation.

There are four types of risk mitigation strategies one could use, namely, risk avoidance, acceptance, transference, and risk control or reduction. While avoidance and acceptance indicate straightforward actions to take, we will discuss *transference* in Chapter 10.

In this chapter, we will focus on risk control or reduction with systematic methodology. This method will also help in establishing quantitative measures which can then help in assessing the progress of risk mitigation activities.

Core principles of AI risk management

As common as it could be for any risk management, in the case of AI risks, three core principles would serve as a foundation. These principles are essential for the effective management of risks:

© Anand Tamboli 2019
A. Tamboli, *Keeping Your AI Under Control*,
https://doi.org/10.1007/978-1-4842-5467-7_7

Identifying risks: Using a systematic approach, one must determine the most severe risks. It is a highly important principle because, without identification, there cannot be any control or management. People have blind spots, and it is our inherent limitation. To avoid them, a good cross section of the team is necessary that will work on risk identification process. The *pre-mortem* analysis method discussed later in this chapter elaborates this principle.

Being comprehensive: As I mentioned previously, AI systems are not one-point solutions. They often involve several systems, and the final system is pretty much system of systems. Which means, to be effective, merely covering AI solution is not going to be helpful. You will have to be comprehensive in risk identification process. Without being comprehensive in this approach, risks from other parts of the system can fall through the cracks and wipe out all the efforts.

Being specific to your use case: It is not difficult to design and implement AI solutions responsibly (i.e., responsible AI); however, to do that it is highly prerequisite to understand risks better and specific to your use case. Broad-based risk identification and management approach will not be helpful. At the same time, establishing a control system across the value chain is also quite important, that is, implementing controls for initial analysis, system design, solution development, implementation, and its ongoing use.

Good things only happen when planned; bad things happen on their own.

—Philip B. Crosby[1]

Frequent contributors to the AI risk

After witnessing several AI failures, and additionally studying many other cases where AI deployments have entirely (or partially) failed, I found a typical pattern of risk contributors. Seven issues existed in almost every case.

While a systematic risk identification would follow, these issues are almost always present and, therefore, do not need any further analysis for fixing:

1. *Flawed hypothesis:* Starting with a wrong assumption often means you are most likely working on the wrong problem, and continuing to do that means you end up retrofitting your solution which will enter in negative spiral of endless fixes.

---

[1] *Quality is Free* (McGraw-Hill, 1979).

2. ***Unattended near misses:*** Quite often minor issues are missed, neglected, or unattended, which pile up over time. Once in a while, drifting from the goal may not cause significant harm, but if you do that too many times, you may end up far away from where you wanted to be.

3. ***Bad data:*** This point does not need any explanation. Incorrect, incomplete, or inaccurate data has been constant caustic problem for several systems and processes in the past.

4. ***Concept drift:*** Performance degradation can happen on account of poorly designed models or due to natural changes in the ecosystem in which AI solution functioning. Either way, concept drift needs continuous monitoring and updates to the machine learning models.

5. ***Poor design:*** There have been several cases, where an elaborate model gets developed to incorporate as many factors (or features) as possible to achieve prediction accuracy. However, when the same model gets deployed in a real-life situation, it fails to meet real-life expectations. The solution itself may be accurate, but it doesn't work in stringent field conditions or expected standards.

6. ***Human-system interactions:*** Interaction points, whether between human and system or one system and other, are always weak links. Failures are highly likely.

7. ***Ingress malfunctions:*** Any input channel to the system, whether it is through direct AI solution or some other upstream/downstream system, can be vulnerable to failures. Moreover, since these inputs eventually flow through more extensive AI system, they carry significant risks.

Pre-mortem analysis

The term project pre-mortem first appeared in the HBR article written by Gary Klein in September 2007. As Gary writes, "A premortem is the hypothetical opposite of a postmortem."[2]

Generally speaking, it is a project management strategy in which a project team imagines a project failure and works backward to determine what potentially could lead to that failure. This working is then used to handle to risks up front.

---

[2] https://hbr.org/2007/09/performing-a-project-premortem

However, in the risk management context, I am going to use this (*pre-mortem*) term interchangeably to represent more sophisticated and engineering-oriented methodology, known as *failure mode and effects analysis*, that is, FMEA.

Critical parts of the pre-mortem analysis

Pre-mortem analysis (or FMEA) is typically done by a cross-functional team of subject matter experts (SMEs). A better format to conduct this exercise is in the form of a workshop.

During the workshop, the team thoroughly analyzes the design and process are implemented or changed. The primary objective of this activity is to find weaknesses and risks associated with every aspect of the product or process or both. Once you identify these risks, take actions to control and mitigate them, and verify that everything is in control.

Pre-mortem analysis record has 16 columns to it as explained here:

1. **Process step or system function:** This column briefly outlines the function, step, or an item you're analyzing. In a multistep process or a multi-function system, there would be several rows, each outlining those steps.

2. **Failure mode:** For each step listed in column # 1, you can identify one or more failure modes. It is an answer to the question—in what ways can the process or solution may fail? Or what can go wrong?

3. **Failure effects:** In the case of failure, as identified in column # 2, what are its effects? How can the failure affect key process measures, product specifications, or customer requirements or customer experience?

4. **Severity:** This column lists severity rating of each of the failure listed in column # 2. Use the failure effects listed in column # 3 to determine the rating. The typical scale of severity is 0 to 10; 0 being the least severe while 10 is the most severe consequence(s).

5. **Root cause:** For each failure listed in column # 2, root-cause analysis is done to find an answer to the question—what will cause this step of function to go wrong?

6. **Occurrence:** This column is another rating that is based on the frequency of failure. How frequently is each of these failures, as listed in column # 2, are likely to occur? Occurrence is ranked on a scale of 1 to 10, where 1 is a low occurrence and 10 is high or frequent occurrence.

7. **Controls:** An answer to the question—what controls are in place currently to prevent potential failure as per column # 2? What controls are in place to detect the occurrence of a fault, if any?

8. **Detection:** Another rating column where ease of detection of each failure is assessed. Typical questions to ask are—how easy is it to detect each of the potential failures? What is the likelihood that you can discover these potential failures promptly, or before they reach the customers? Detection is ranked on a scale of 10 to 1 (note reversal of the scale). Here rating of 1 means easily and quickly detectable failure, whereas 10 means unlikely and very late detection of failure. Detection late often means more problematic situation, and therefore the rating for late detection is higher.

9. **RPN (Risk Priority Number):** The risk priority is determined by multiplying all three ratings from column # 4, 6, and 8. So, RPN = *Severity x Occurrence x Detection*. Thus, high RPN would indicate a high-risk process step or solution function (as in column # 1). Accordingly, steps or functions with higher RPN warrant immediate attention for fixing.

10. **Recommended actions:** In this column, SMEs would recommend one or more actions to handle the risks identified. These actions may be directed toward reducing the severity or reducing the chances of failure occurrence, or to improving the detection level, or maybe all of the above.

11. **Action owner, target date, and so on:** This column is essential from project management point of view as well as for tracking. Each recommended action can be assigned to a specific owner and carried out before target date to contain the risks.

12. **Actions taken:** This column lists all the actions taken, recommended or otherwise, to lower the risk level (RPN) to an acceptable level or lower.

13. **New severity:** Once the actions listed in column # 12 are complete, the same exercise must be repeated to arrive at new level of severity.

14. **New occurrence:** Depending upon the completed actions, the occurrence must have changed, so this column has new occurrence rating.

15. **New detection:** Due to risk mitigation actions, detection must have changed, too; register it in this column.

16. **New RPN:** Due to change in severity, occurrence, and detection ratings, risk level would have changed. New RPN is calculated in the same way, *Severity x Occurrence x Detection*, and recorded in this column.

More about ratings

Several risk analysis methodologies often recommend only two rating evaluations, that is, severity and occurrence. However, in the case of pre-mortem analysis, we are using third rating—detection.

Early detection of the problem can often enable you to contain the risks before becoming significant and out of control. This way, you can either fix the system immediately or may invoke system-wide control measures to remain more alert. Either way, being able to detect failures quickly and efficiently is an advantage in complex systems like AI.

In the case of severity and occurrence ratings, the scale of 1 to 10 does not change based on the type of solution or industry as these are generally defined scales.

In implementing pre-mortem analysis, you must take a pragmatic approach and choose the scale as appropriate. Just make sure that you are consistent in your definitions throughout the pre-mortem exercise.

While conducting a pre-mortem workshop, participants must set and agree on rankings criteria up front and then for the severity, occurrence, and detection level for each of the failure modes.

Using the output of the analysis

The output of pre-mortem analysis is only useful if you use it.

Each process step or system function would have one or more RPN values associated with it. The higher the RPN, the riskier the step is. During the pre-mortem exercise, the team must decide a threshold RPN value. This way, for all the steps where RPN is above the threshold, risk mitigation and control plan become mandatory, whereas for RPNs below threshold may be addressed later as their priority would be lower.

Ideally, you should be addressing all the practical steps wherever RPN is non-zero. However, it is not always possible due to resource limitations.

One of the ways you can reduce RPN is by reducing the severity of the failure mode. Typically, reducing severity often needs functional changes in process steps or the solution itself. Additionally, the occurrence can be controlled by addition of specific control measures such as human in the loop or maker-checker mechanisms.

However, if it is not possible to reduce the severity or occurrence, then by implementing control systems, you can contain the failures. Control systems can help in either detection of causes of unwanted events before the consequence occurring or the detection of root causes of unwanted failures that the team can then avoid altogether.

Having risks quantified and visible also enables you to have plans in place to act quickly and appropriately in case of failures and thus reduces the exposure to more failures or adverse consequences.

It is possible that during the pre-mortem exercise, the team will discover many failure modes or root causes that were not part of the designed controls or test cases/procedures. It is crucial that the test and control plan is always impacted by the results of this analysis. Ideally, you must include test and control team members for pre-mortem analysis exercise.

A common problem I've seen in this exercise is difficulty or failure to get to the root cause of anticipated failure, and this is where SMEs should lean in. If you do not identify root causes correctly or do it poorly, then your follow-up actions would not yield proper results.

Another problem I've seen is the lack of follow-up to ensure that recommended actions are executed and the resulting RPN is lowered to an acceptable level. Doing effective follow-through is project management function and needs diligent execution to ensure that pre-mortem analysis reaches its logical conclusion.

Pre-mortem analysis workshops can be time-consuming at times. Due to high time demand, it may become challenging to get sufficient participation of SMEs. The key is to get the people who are knowledgeable and experienced about potential failures and their resolutions showing up at these workshops. SME attendance often needs management support, and facilitators need to ensure that this support is garnered.

You can read more about FMEA find quality information on the Internet.[3]

---

[3] https://accendoreliability.com/fmea-resources/

Sector-specific considerations

In pre-mortem analysis, Severity-Occurrence-Detection (SOD) ratings range between 1 and 10. However, the weights assigned to each of the rating value are subjective. It is possible that in the same industry two different companies could come up with slightly different ratings for the same failure mode.

To avoid too much subjectivity and confusion, some level of standardization or rating scale could be helpful. However, that would be only necessary when you're benchmarking two or more products from different vendors in the same industry. If this has to be used only for internal purposes, subjectivity won't matter much, since relative weights and essential would still be preserved within the risks and action items.

Nonetheless, when considering control or action plans for identified risks, sector-specific approaches could be (and should be) different.

Any failure risk can be controlled by either reducing severity (S) or by lowering the chances of occurrence (O) or by improving detection levels (D). If this were to be done in the banking sector, while enhancing S and O ratings, D ratings might need additional focus for improvement. Given the volume of transactions that financial sector carries out every day, severity of failure could be high due to widespread impact, but if the severity can't be controlled beyond a point, detecting it early to fix would be highly necessary.

In the case of healthcare sector though, severity itself should be lower as the detection may likely result in fixing a problem but wouldn't necessarily reverse the impact. For example, if AI prediction or solution results into incorrect prognosis and thereby change in medicine, early detection of this problem may result in stopping the activity per se. However, it won't be able to revert the issues caused by having this failure in the first place.

Similarly, in transportation scenario, especially for autonomous cars, detecting that a car's mechanism has failed as an after the fact is less useful, since the accident would already have happened. Reducing severity and occurrence in those cases is a more acceptable course of action.

Generally speaking, you should focus on improving detection, if the impact of failure can be reversed in due course of time or there is enough time available between system's outcomes and see the full effect on the end user. If even having one failure means significant damage to you, then severity must be reduced, and occurrence must be reduced too.

Severity and occurrence improvement are prevention focused, whereas detection improvement is fixing (cure) focused. If your industry believes that prevention is better than cure, then work on to reduce severity and lower the occurrence of failures. If your industry is comfortable with fixes after the fact, then detection must be improved.

However, in my view, it is better to address all three factors and ensure that robust risk management is in place.

Conclusion

Pearl Zhu, in her book, *Digitizing Boardroom*[4] says, "Sense and deal with problems in their smallest state, before they grow bigger and become fatal."

Managing risks systematically and pragmatically is the key to handle AI risks. The problem with AI risks is that they are highly scalable and can quickly grow out of control due to power of automation and sheer capacity of AI to execute tasks.

Moreover, subjectivity in risk management is a myth. If you can't quantify the risk, you can't measure it; and if you can't measure it, then you can't improve or control it. The systematic approach outlined here will help you to quantify your risks and understand them better while maintaining the context of your use case.

You can develop and implement AI solutions responsibly; if you understand risks…understand them better and specific to your use case!

---

[4] Self-published, 2018

Teach Meticulously and Test Rigorously

If you want your system to perform better, you ought to teach it better; and, how do you verify if you have trained it well? By testing it better!

You will find several resources that cover the AI from a development perspective; however, there are hardly any, which include testing AI system as such.

AI testing is quite different compared to the typical software testing. The challenges and risks any AI system pose can make it even more critical to have a rigorous process and test mechanism in place.

Majority of the AI solutions learn in two different stages. The first learning happens while working with controlled datasets and formulating base models. The second learning occurs on the go or periodically with the help of user interactions in the form of feedback.

© Anand Tamboli 2019
A. Tamboli, *Keeping Your AI Under Control*,
https://doi.org/10.1007/978-1-4842-5467-7_8

Let us discuss how to navigate through teaching and testing of an AI system to avoid risks and set it up for continuous improvements.

How AI learns

First, let us understand the learning stages and methods of learning. Doing so enables us in appreciating pros and cons of each and also helps in understanding potential loopholes. Each of that may result in risky outputs.

A sophisticated AI system will usually follow a two-stage learning mechanism, whereas the simple AI system may have only one formative stage. The two stages in which AI learns are

- Training
- Feedback

The training stage is where basic machine learning models are formed for the first time, and the system is trained using those models. The accuracy of these models is highly dependent on the training dataset.

On the contrary, when the AI system is deployed, some (or all) data can be fed back to the system for continuous learning, which we term as feedback stage learning. This stage is relatively vulnerable to various ongoing risks. Mainly because no matter how accurate you develop your base models in the past, new data (coming through feedback system) can cause these models to readjust and thus become better or worse.

However, a lot would also depend on how the machine is learning in each stage.

Types of machine learning

There are three types of machine learning:

1. **Supervised** machine learning is usually task-driven, where the objective is to teach and train the algorithm for prediction. A commonly quoted example of this type of goal would be *cat or no cat* identification system or say predicting tomorrow's weather. Spam classification and face recognition are some other examples of this type of learning outcomes.

   This algorithm consists of a target or an outcome variable (also known as dependent variable) which the system needs to predict from a given set of predictors (aka independent variables). Using these variables, we can generate a function that maps inputs to desired outputs.

We continue the training process until the model achieves a desired level of accuracy on the training data.

2. ***Unsupervised*** machine learning is highly data-driven, where the objective is to identify patterns and clusters from the data. With unlabeled data input, the algorithm is expected to spit out grouped or clustered datasets for further classification, mostly by the supervisor.

   Here, we do not have any target or outcome variable to predict or estimate. The output is mainly clustered or grouped and can be used as a recommendation or request for classification. Consider YouTube or Netflix recommendations, which monitor your video watching pattern, cluster similar videos together and recommend them to you.

3. ***Reinforced*** learning in simple terms is *learning from mistakes*. While supervised and unsupervised learning predominantly indicates the presence or absence of feature labels, this one is different.

   With reinforced learning mechanism, we let the algorithm make prediction then reward it if the prediction is correct and penalize it if not. Based on the rewards and penalties, over time, the algorithm adjusts its performance to make fewer mistakes and continues to predict. When the model achieves a reasonable level of accuracy, the reinforced training stops.

Five mistakes to avoid when training

Meticulous teaching is the fundamental requirement for having an excellent performance consistently. However, there are a few caustic mistakes to which traditional (and contemporary) data analytics or statistical methods are susceptible. The collective calling for such issues is often known as *garbage in, garbage out*.

Not having enough data to train

You may ask, "How much data does one need to train the AI system effectively?"

I'd say, "It depends!".

It may be a sour answer, especially if you are at the pointy end of your machine learning stage. Nevertheless, it indeed depends on the complexity of your problem at hand as well as the complexity of the algorithm you plan to use. Either way, the best way would be to use empirical investigation and arrive at an optimal number.

You may want to use standard sampling methods in the collection of required data and may wish to use standard sample size calculators as used in standard statistical analysis tools. However, due to the nature of machine learning algorithms, the amount of data is often insufficient. You most likely would need more than what a standard sample size calculation formula tells you.

Having more data may not be an as big problem as having it less would be. You have to make sure that there is enough data to reasonably capture the relationship that might exist within input parameters (aka features) and between input and output.

You may also use your domain expertise to reasonably assess how much data is enough to exhibit a full cycle of your business problem. It should cover all the possible seasonality and variations.

The model developed with the help of this data will only be as good as the data you have or provide for training, so make sure that it is adequately available. If you feel that the data is not enough, which may be a rare scenario in the current big-data world, don't rush; wait until you get enough of it.

Not cleaning and validating the dataset

Too much data is of no use if it is of poor quality and can mean one or more of the following three things:

1. **Data has noise**, that is, there is too much conflicting and misleading information. Confounding variables or parameters are present and essential variables are missing. Cleaning this type of data needs additional data points, because the current set is unusable and hence not enough.

2. **It is dirty data**, that is, several values are missing (though parameters are available), or the data has inconsistencies, errors, and mix of numerical or categorical values in the same column. This type of data needs careful manual cleaning by subject matter experts and may often need re-validation. Depending on the resource availability, you may find it easier to obtain additional data instead of cleaning dirty data.

3. **Inadequate or sparse data** is the scenario where very few data points have actual values and a significant part of the dataset is full of *nulls* or zeroes.

The type of issues present within the dataset is often not clear from the dataset itself, which is why I always recommend exploratory analysis and visualization to be applied at the outset. Doing this first pass not only gives you a level of confidence in data quality but also can tell you if there is something amiss.

Based on the visual representation, an interesting question would be—do you see what you expected to see?

If the answer is "No," then your data may be of poor quality and needs cleaning.

If the answer is "Yes," it might be useful in finding some preliminary insights. This validation of dataset is essential to proceed, and you should never miss it.

Not having enough spread in data

Having a large amount of data is not always a good thing unless it can represent all the possible use cases or scenarios. If the data is missing variety, then it can lead to problems in the future—you increase the chances of losing on low-frequency high-risk scenarios.

For traditional predictive analysis, there is a point of diminishing returns as you obtain more and more data for training. Your data science team can often spot this point empirically.

However, since machine learning is an inductive process, your base model can only cover what it has seen in the data. So, if you miss on long tail, aka edge cases, they will not be supported by your model. It merely means your AI will fail when that scenario occurs. That is the only and the most crucial reason why your training data should have enough spread to represent the real population.

Ignoring near misses and overrides

During initial training, it is hard to identify near misses and disregarded data points. However, in a continuous learning loop with feedback, it becomes highly essential to pay close attention to near misses and human or machine overrides.

When you deploy your AI system for the first time, it has an only base model that governs the performance of an AI. However, as system operation continues, the feedback loop feeds live data, and system starts to adjust, either live or regularly.

If the model has missed to correctly predict or calculate any output just by a bit and thereby the decision has changed, it would be a near miss. For example, in the case of a loan approval system, if 88.5% score means "*loan approved*" and 88.6% results in "*loan declined,*" then this scenario is a near miss. From a technical and pure statistical point of view, it is correct; however, from a real-life perspective, a margin of error may play a significant role. If contested by the affected party, such as loan applicant, chances of change in a decision are higher. Therefore, these type of data points are of particular interest, and you should not ignore them.

The same applies when a human operator is supervising AI system's output and can decide to override it. Human operator overriding output of an AI should always be treated as a special-case scenario, and you must feed it back to the training model. Each of these scenarios either highlights inadequacies in the base model or provide new situations that never existed before. Ignoring overrides can degrade the model performance over time.

Conflating correlation and causation

In statistics, we often say, "correlation does not imply causation." It usually refers to the inability in legitimately deducing a cause and effect relationship between input variables and output. The resulting conclusion still may not be incorrect or false, but the failure to establish this relationship is often an indicator of lurking problem.

On similar terms, the predictive power of your model does not necessarily imply that you have established an exact cause and effect relationship in your model. Your model may very well be conflating correlation of input parameters and predicting output based on that.

You may think that "As long as it works, it shouldn't matter." However, the distinction matters since many machine learning algorithms pick upon parameters simply because there is a high correlation. Determining causality based on correlations can be very tricky and can potentially lead to contradictory conclusions. It would be much better to be able to prove that there truly is a causal relationship.

However, these days, developers and data scientists are merely relying on statistical patterns. Many of them fail to recognize that those patterns are only correlations among vast amounts of data rather than causative truths or natural laws, which govern the real world.

So, how do you deal with conflation?

Try this—during initial training and model building, soon after you find a correlation, don't conclude too quickly. Take time to find other underlying factors, find the hidden factors, and verify if they are correct and then only conclude.

Complementing training with testing

If you have to trust someone with their performance, there is one of the two ways to do it. Train them effectively such that their performance is guaranteed. If you suspect training, even by a bit, then test them rigorously to ensure a better performance. Moreover, if you do both, that is, train meticulously and test rigorously, you can be confident about the performance, and it forms a better basis for trust.

So, while you would focus on better training of your AI system, it would also make sense to complement that training with better testing rigor.

Better training would imply that machine learning has not resulted in overfitting or force-fitting of models. However, if you suspect that this may have occurred, you should craft your testing to capture it. You can easily do that with brand new datasets, albeit relevant to the application, where the dataset represents real-life scenarios.

If you suspect that training was not adequate and may not have covered all the possible scenarios, including risky ones, then create tests that can put your AI system under these conditions and verify that system performs as expected.

Design your tests to catch potential pitfalls that may occur during the training. These tests should also reflect the conditions where your AI system must perform and how would you ascertain. If your AI passes those tests, then you know that it is good enough.

You will often come across black-boxed AI systems or at least some components of it, which would be like that. Since you would be limited to understand the inner working of those systems or sub-systems, your only allies would be meticulous training and then rigorous testing. Testing should reveal any potential threats and opportunities, and if all goes well, you would be reasonably confident about the black box's functioning.

In a nutshell, take rigorous statistical testing seriously, and you would be good to go (in production).

AI testing is not software testing

Yes, AI system testing is not the same as software testing.

Traditional software testing is entirely black-and-white activity. You have a fixed test case (with expected response), where you test it with designed input. If you find a problem, you note the scenario and fix it.

However, with machine learning, it plays differently because, in most of the cases, it will be almost impossible to retrace decision-making of the algorithm. Moreover, testing the AI system would often incite a lot of (rather far too many) scenarios, which is uncommon in traditional testing.

In traditional software testing, when we act, we would expect the same outcome each time we repeat it. With AI, it may not reproduce the same result 100% of the times.

Also, since you cannot test case-by-case basis, you may need to throw a huge number of scenarios at the system and see what you get. If the proportion of correct output is reasonably high, then you can be confident that it will work (most of the times). Much like our school exams, where we may not get 100% marks, but having them beyond specific percentage means, we are good enough to ascend to the next level.

The cardinal sequence for AI testing

Any testing, whether it is traditional software testing or AI testing, should start with the premise—*there is a problem (or bug) until proven otherwise by way of tests*.

A typical software testing sequence is quite logical and almost immutable due to its limited nature.

However, in the case of AI testing, change in sequence is possible, and this change has a full potential to make or break the system performance. If you do not test AI in proper sequence, it can increase the risk exposure of the business significantly:

1. ***First decide what to test, that is, define your population of interest***

   You may not be required to define test cases or individual test scenarios; however, you will need to specify how the test dataset would look. You will need to define overall characteristics of the test dataset, a broad collection of instances, and potentially, where to get it? These have to be representative ones, such that if your system demonstrates a good performance on their testing, you will approve it for production.

2. ***Define how would you test, that is, testing criteria***

   Testing criteria should be reasonably straightforward to define—how would you test your AI system. What would you do to test it?

3. **Decide what will you accept, that is, quantified acceptance criteria**

    Define the minimum acceptable performance that you will be ready to sign off. Describe, quantitatively, what level of performance is good enough for you. Define, what does good enough means to you; and once you do it—stick to it!

4. **Complete AI testing**

    Once the preceding three steps are complete, it is time to test. Do all the testing and accept or reject the system based on the criteria set earlier. Avoid bargaining with the acceptance criteria once you fix it. As humans, we tend to fall in love with what we have poured time and effort into, even if it is all junk. In economic jargon, we call it as the endowment effect. You should stay away from it!

In short, decide what to test—how to test—what to accept or reject—then test. Stick to this sequence, and you would be fine.

If you find yourself deciding or negotiating acceptance criteria after the testing, then you're doing it wrong. Avoid exposing your business to high financial and reputational risks; these are irreversible.

No matter how much you test, do not test excessively, with high passing bar lest you could miss out on a good solution.

Testing for concept drift

A concept in "concept drift" refers to the unknown and hidden relationship between inputs and output variables. The change in data can occur due to various real-life scenarios.

Training and feedback can mutate dynamically. These transmutations are very diverse and affect the reliability of the base model. The problem of change in data over time and thereby affecting statically programmed (or assumed) relationships is a common occurrence for several other real-life scenarios of machine learning. The technical term, in the field of machine learning, for this scenario is "concept drift."[1]

---

[1] https://medium.com/tomorrow-plus-plus/handling-concept-drift-at-the-edge-of-sensor-network-37c2e9e9e508

It is difficult to identify a scenario in which concept drift may occur, especially in the first stage of training, that is, when you're developing the base model. Therefore, it is a better choice to safely assume that concept drift will occur and test the model for potential drift.

There are various statistical methods, which you can use for detecting the concept drift early on in the process. While we will not discuss those analytical methods for concept drift testing, a simple Internet search with the term "*testing for concept drift*" can give you enough material to work.

In a nutshell—you should make sure that you are testing your model before deployment as well as on an ongoing basis for concept drift.

Testing for interaction issues

The AI systems are not stand-alone, and they often interact with more than one other system and humans too. At each interaction point, there lies a chance of failure or subpar performance. I always say that every link that joins two heterogeneous systems is a weak link.

To uncover these weaknesses and be able to fix them, you need to test system interactions at all those points.

There are two objectives to test human and AI system interaction. First is to validate usability issues that are mostly UI and UX related. Second is to prove that performance in a multi-user environment.

The first part is usually easy to achieve with standard UI/UX testing. However, for the second part, things can get a little complex. As mentioned in previous sections, AI performance testing is not a point-based scenario testing. Instead, you feed it with the bulk of inputs and expect a reasonable proportion of them to be in acceptable limits, which adds some level of complexity in your interaction testing. Then come the repeatability and reproducibility of each of the user.

In statistical analysis, we call it *Gauge R&R*, that is, gauge repeatability and reproducibility. You would reasonably expect a human operator to demonstrate the same performance repeatedly for any given set of scenarios. You would also expect them to reproduce the same outcome each time you repeat those scenarios.

When you mix this with an AI system that works on acceptable proportional outcomes, things drastically change. Here you are dealing with system variation (though acceptable) coupled with human variation (again within acceptable limits). Both these variations, when applied to several systems and several humans in the loop, mean that the whole operational machinery is now exhibiting significant variance in performance. Despite being utterly acceptable to some extent, this poses a considerable risk, mainly due to the possibility that a minor variation in any of the actors' performance can result in poor overall performance.

That's the reason why part of your AI system testing should also focus on gauge R&R for all the humans involved in the system's loop. Of course, when the person changes, you must do over this with a new contributor. It can be a painful exercise. However, to maintain a high level of performance of your AI system, it is a small price to pay.

The same level of challenge you would face when testing interactions between several software systems, for instance, your AI interacting with say CRM system or ERP or trouble ticket system. It may also be the case that your AI is interacting with systems outside your control, such as customer's or vendor's systems. It is crucial that you test those interactions with equal rigor and diligence to uncover any systemic or repeatability and reproducibility issues.

The key is not only to test these interactions before deployment but keep doing this testing on an ongoing and regular basis. With such an elaborate system of systems in place, every time a contributor (human or system) changes, the dynamics change, and so must your tests.

It can be the basis for your trust

As technology is improving day by day, it is placing powerful tools in the hands of people, who do not understand how they work. It is creating a significant business as well as societal risks.

Developers and data scientists are increasingly getting detached from an understanding of intricacies of the tools they are using and systems they're creating.

The AI system means a black box is becoming a commonly accepted rhetoric, and the only surefire way to trust this black box is going to be—training it meticulously and testing it rigorously!

The principal objective of rigorous testing is to become confident and increase trust in future performances.

Mitigation

Minimizing impact of risks

AI Supervision with a Red Team

How do you stop an attacker? By thinking and planning like one!

We have a supervised machine learning as a concept, but supervising an AI itself is not deeply thought out. In this chapter, we focus on ongoing controls over the AI solution. We will discuss what does it involve supervising an AI and elaborate mainly the concept of a red team in the context of AI solutions.

What is a red team?

In the military parlance, "red team" is the team which uses their skills to imitate the attack and techniques that potential enemies might use. The other group, which will act as a defense, is called a "blue team."

Cybersecurity has picked up these terms for the use, where they signify the same functionality.

A red team is an independent group of people that challenges an organization to improve its defenses and effectiveness by acting as an adversary or attacker. In layman's terms, it is a form of ethical hacking, a way to test how well an organization would do in the face of a real attack.

© Anand Tamboli 2019
A. Tamboli, *Keeping Your AI Under Control*,
https://doi.org/10.1007/978-1-4842-5467-7_9

If effectively implemented, red teams can expose vulnerabilities and risks not only in your technology infrastructure but also in people and physical assets.

You would often hear a saying that "the best offence is a good defense." Having a red team is the right step in putting up this good defense. Whether it is a complex AI solution or merely basic technology solution, red teaming can give a competitive edge to the business.

A comprehensive red team exercise may involve penetration testing (also known as pen testing), social engineering, and physical intrusion. Either all of these could be carried out, or a combination of those in the right manner as the red team would see fit to expose vulnerabilities.

If there exists a red team, then there also must exist blue team. This assumption stems from the premise that system developments are done in-house only. However, with AI systems, this can change, and your actual blue team maybe your technology vendor.

Supervising an AI is necessary

So far, the businesses have (rarely) used red teams only for the typical endeavors of cybersecurity. However, using the same for monitoring an AI system is not yet there. Most like likely because many businesses still see the AI system as just another IT system.

However, with the advent of sophisticated AI systems, boundaries between mere IT systems and other business processes are becoming unclear and are slowly diminishing. This proliferation of systems means your AI system is no more just a piece of software, but it is an integral part of your business process. A quick test can help in making a clear distinction between a system that is an essential part or otherwise. Ask yourself, "If the system were stopped completely for any reason, can your business process still function as usual?" If the answer is negative, then you can be sure that your systems are pervasive enough.

Having a functional red team also means that you can identify asymptotic situations that can result in catastrophic behavior of your AI system. Being able to identify these situations is essential from building a reliable operation in the long term. Moreover, one shouldn't discount a capable adversary who may be fixated on a goal to bring your AI system down and cause some harm.

Since most of the AI solutions are appearing as black boxes, it is becoming more relevant to supervise them regularly. Not monitoring these systems could mean that things would go wrong without getting noticed. It would be a significant problem if you see issues somewhat later and would be a huge trouble if you never notice them until severe consequences occur.

Practically, we train machine learning systems on a fixed dataset. However, in reality, we almost always subject these systems to the data outside of their training distribution (long tail as some would call it). How can you be sure that AI systems work robustly on the real data?

Wait and watch cannot be an approach here. Some machine learning models you could train on the fly, mostly as unsupervised learning mechanisms. You can feed new training inputs back to the system, and then the system keeps optimizing continuously. However, while this might be good enough sometimes, at other times, failure can be disastrous.

Customers may, at times, tolerate a hundred useless product recommendations. However, they certainly cannot tolerate a single mistake that can be hazardous or causes them to lose a significant amount of money. Business users are no different here.

Training an AI system to fall back on human intervention can be one of the strategies. However, some may often question its effectiveness as it is counterintuitive to the existence of an AI in the first place.

Relying solely on fine-tuning of machine learning models to produce more and more accurate output is not enough. It will never give you a clear picture of the actual risk exposure.

Why not use adversarial AI testing?

A short answer to this would be: because adversarial AI testing is simply a technology testing method, where you would play one AI against another. However, the red team concept is cross-domain. Not just cross-domain, it is cross-sectional in nature.

If it was only adversarial testing from a technology standpoint, a red team may try to find an image that could throw off image recognition AI systems' output. The same result can be effectively achieved with automated mechanisms too.

However, a real red team would not stop at just finding an image to break the system. It would go above and beyond that to create or doctor an image to make it happen, which is where adversarial AI testing may fall short. The GAN (generative adversarial network) systems could prove to be narrower in that context, but the red team would be working much more broadly.

The ultimate goal of red teams is to produce problematic inputs or situations that are natural and have a nonzero probability.

Businesses have always siloed physical and technical securities. The people who oversee IT systems—networks, applications, and alike—are not the same people who manage physical security such as cameras, motion sensors, and locks. You may have the best cyber defenses in place. However, if your physical safety is weak, then eventually, someone will make their way in your systems through that route. Moreover, the opposite may also be possible.

Therefore, in the world, with highly connected devices and systems, there is a need to break the silo.

Only AI supervision can expose this

Several machine learning techniques that feed to the AI system rely on finding correlation within a supplied dataset. Whether it is training dataset or ongoing feedback dataset, mostly they would be treated in the same manner. Supervising your AI system can be beneficial at many levels.

Highlighting statistical bloopers

By and large, these analytical techniques depend on summary statistics of the dataset. This approach can be a significant problem because multiple datasets with different characteristics can have the same summary statistics. In February 1973, F. J. Anscombe published a journal article where he says, "A computer should make both, calculations and graphs. Both sorts of output should be studied; each will contribute to understanding."[1]

Inspired by Anscombe's Quartet and the Datasaurus, in August 2016, Alberto Cairo created the Datasaurus[2] dataset. This dataset urges people to "never trust summary statistics alone; always visualize your data," since, while the data exhibits normal-seeming statistics, plotting the data reveals a picture of a dinosaur.

Autodesk, in 2017 published[3] a detailed paper titled "Same Stats, Different Graphs: Generating Datasets with Varied Appearance and Identical Statistics through Simulated Annealing," on these new characteristics of datasets.

As you go through all these references, you would realize that from an attacker's perspective, this is a boon. Also, in real life, the occurrence of these types of scenarios is possible. Without having a capable red team and supervising AI regularly, these issues could turn into threats and may go undetected for too long.

---

[1] Anscombe, F. J. "Graphs in Statistical Analysis." The American Statistician, vol. 27, no. 1, 1973, pp. 17–21. JSTOR, www.jstor.org/stable/2682899.www.jstor.org/stable/2682899?origin=crossref&seq=1#page_scan_tab_contents

[2] www.thefunctionalart.com/2016/08/download-datasaurus-never-trust-summary.html.

[3] www.autodeskresearch.com/publications/samestats

Detecting concept drift

In addition to the statistical bloopers, supervising AI can help you identify a phenomenon known as "concept drift." A concept in "concept drift" refers to the unknown and hidden relationship between inputs and output variables. The change in data can occur due to various real-life scenarios.

More so, this problem of change in data over time and thereby affecting statically programmed (or assumed) underlying relationships is a common occurrence for several other real-life scenarios of machine learning. The technical term, in the field of machine learning, for this scenario is "concept drift."[4]

You may ask, "Why is this a problem?".

For an utterly static use case, this (concept drift) is not a problem at all. However, in several use cases, the relationship between input parameters (or features) and output characteristics changes over time. If your machine learning model did assume data patterns to be static, there would be a problem in the future. This drift cannot be predicted as the changes in the environment or affecting factors could be random or fabricated. Either way, without supervising AI system, you would never be able to uncover them.

One of the significant challenges in dealing with concept drift is to detect when this drift occurs. There I suggest one of two ways to handle that.

When you finalize a machine learning model for deployment, record its baseline performance parameters such as accuracy, skill level, and others. When you deploy the model, periodically monitor these parameters for change, that is, supervise AI regularly. If you see the difference in parameters is significant, it could be indicative of potential concept drift, and you should take action to fix it.

The other way to handle it is to assume that drift will occur and, therefore, periodically update the model.

It is unrealistic to expect that data distributions stay stable over a long period. The perfect world assumptions in machine learning do not work in most of the cases due to changes in the data over time. This limitation is a growing problem as the use of these tools and methods increases.

Acknowledging that AI or machine learning may remain inside a black box for a long time and still they would evolve continuously is the key to fixing it.

---

[4]https://medium.com/tomorrow-plus-plus/handling-concept-drift-at-the-edge-of-sensor-network-37c2e9e9e508

Finding what's inside the black box

AI systems being the black box is going to be a pertinent and ongoing problem. Regardless of the influx of sea of trust issues and questions, it will be near impossible to get access to the internal logic of AI systems.

However, this black-box problem can be handled slightly differently with the help of red teams in operation and supervising AI.

This approach, however, needs a careful designing of experiments and scenarios by testing and red teams and then executing them systematically. When done correctly, users or organizations can reasonably estimate what is going on inside the black box and then act accordingly.

Remember that this is true for external parties too, who would try to manipulate your AI systems in malicious ways. So this is why AI supervision can play an essential role in identifying these risks up front or as early as possible.

Building your red team

Ideally, a red team needs at least two people to be effective, though many ranges from two to five. If you are a large company, you might need 15+ members in the team to work on several fronts.

Depending upon the type of AI you are deploying, your red team's composition might change with different skill sets. The structure is necessary to maximize the team's effectiveness.

Typically, you would need physical security experts, such as the ones who understand and can deal with physical locks, door codes, and other aspects. You would also need a social engineer to phish out information through emails, phone calls, social media, and other such options. Also, most importantly, you would need a technology expert, preferably a full stack one to exploit hardware and software aspects of your system. These skill set requirements are the minimum. If your application and systems are too complicated, then it will make sense to hire specialists for individual elements and have a mixed team of experts.

Top five red team skills

The most important skill any of the red team members can have is to think as negatively as possible and remain as creative as they can be when executing the job:

- *Creative negative thinking:* It is the core goal to continually find new tools and techniques to invade systems and eventually protect the organization's security. Moreover, showing the flaws in the systems needs a level of negative thinking to counter the inbuilt optimism in an individual.

- *Profound system-wide knowledge:* It is imperative for red team members to have a deep understanding of computer systems, hardware and software alike. They should also be aware of typical design patterns in use.

  Additionally, this knowledge shouldn't be limited to only computer systems. It must span across many systems to involve heterogeneous systems.

- *Penetration testing (aka pen testing):* This is a common and fundamental requirement in the cybersecurity world. Moreover, for red teams, it becomes an essential part and kind of standard procedure.

- *Software and hardware development:* Having this skill means as a red team member, you can envisage and develop tools required to execute your tasks.

  Moreover, knowing how AI systems are designed in the first place means you are likely to know failure points. One of the critical risks AI systems always pose is "logical errors." These errors do not break the system but make them behave in a certain way that is not intended or may cause more damages. If a red team member has experience in software and hardware development, then it is highly likely that they have seen all the typical logical errors. Therefore, they can exploit them to accomplish their job.

- *Social engineering:* This goes without saying as manipulation of people into doing something that can lead the red team to their goal is essential. The people aspect is also one of the failure vectors that the actual attacker would use. Human errors and mistakes are one of the most frequent reasons for cyber problems.

In-house or outsourced

Now the next key question is—should you hire your team members and employ in-house or outsource the red team activity?

We all know that security is one of those aspects where funding is mostly dry. It is tough because the ROI of security initiatives cannot be proven easily. Unless something goes wrong and you prevent it, it is nearly impossible to visualize or imagine. This limitation makes it difficult to convince that the investment in security is worth doing, and you are investing money wisely.

So, a quick answer to an in-house or outsourced question would be—it depends on company size and the budget.

If you have deployed AI systems for long-term objectives, then the in-house red team would be the right choice as they would be engaged continuously. However, that comes with an additional ongoing budget overhead.

On the contrary, if you are unsure about your overall outlook, outsourcing is a better way to start. This way, you can test your business case for in-house hiring in the long run.

From privacy and stricter controls perspective, the in-house red team is highly justifiable. The red team and blue team activities are more like cat-n-mouse games. When done correctly, each round can improve and add to the skill set of both the teams, which in turn would enhance the organization's security.

You can utilize the outsourcing option if you are planning to run a more extensive simulation. If you need specialized help or looking for a specific skill set for particular strategy execution, then also it would make sense.

Objectives of a red team

Primarily, the red team exists to break the AI system and attached pro-cesses by assuming the role of the maleficent entity. The red team should to go beyond just the technology aspect and work on the entire chain that involves the AI system. Doing this can make their efforts more effective as it can then ensure that upstream as well as downstream processes and sys-tems are tested.

A red team should consider the full ecosystem and figure out how a deter-mined threat actor might break it. Instead of just working toward breaking web app or particular technology application, it should combine several attack vectors. These attack vectors could be outside the technology domain, such as social engineering or physical access if needed. It is necessary because although your ultimate goal would be to reduce AI systems' risks, these risks can come from many places and in many forms.

To maximize a red team's value, you should consider a scenario and goal-based exercise.

The red team should get into motion as soon your primary machine training is complete, which applies if you are developing the model in-house. If you are sourcing trained model(s) from an outside vendor(s), then the red team must be activated as soon as sourcing completes.

The primary goal of a red team would be to produce or create a plausible scenario in which the current AI system behavior would be unacceptable and, if possible, catastrophically unacceptable. If the red team succeeds, then you can feedback their scenarios to the machine training team for retraining of the model. However, if the red team does not succeed, then you can be reasonably confident that the trained model will behave reliably in the real-world scenario too.

Carefully staging potentially problematic scenarios and exposing the whole AI system to those situations should be one of the critical objectives. Also, this activity need not be entirely digital in format. The red team can generate these scenarios by any means available and in any formats as they seem plausible in real-life situations.

One of the ways the red team can attempt to fail the AI system is by giving garbage inputs in primary or feedback loop and seeing how it responds. If the system is smart, it will detect the garbage and act accordingly. However, if the system magnifies or operates on the garbage input, you will know that you have work to do. These (garbage) inputs can take a form of training inputs for machine retraining.

Red teams can also work on creating and providing synthetic inputs and see how the system responds. The output then they can use to examine the AI system's internal mechanics. Based on further understanding, synthetic data could be made more authentic to test the system's limits, responses, and overall behavior. Once you identify failure situations, they are easier to fix.

Red teams may not necessarily try to break the systems. Sometimes, they may merely cause a drift in the continuous learning of the system by feeding wrong inputs or modifying parameters and thereby cause it to fail much later. Refer to concept drift phenomena discussed in the earlier section. While concept drift is mostly natural and normal, it can be deliberate and manufactured.

A point where your AI system is taking input from another software or human could be a weak link. A point where the output of an AI system forms an input to another API or ERP system could also be a weak link. By nature, these links are highly vulnerable spots and weak links of the whole value chain.

Every link that joins two heterogeneous systems is a weak link!

Red teams should target and identify all such weak links in the system. These weak links may exist between two software systems or at the boundary of software-hardware or software-human interaction.

A red team is not for testing defenses

The core objective of the red team is not to test your defense capabilities. It is to do anything and everything to break the functional AI system, in as many ways as possible, and by thinking outside the box. Ultimately this should strengthen the whole organization in the process.

Having this broader remit can enable red teams to follow intuitive methodologies for developing a reliable and ongoing learning system for the organization. It is a promising approach to many severe problems in AI systems' control.

However, remember that red teaming is not equivalent to a testing team who work on generating test cases. Test cases usually follow a well-defined failure condition(s), whereas for the red team objective is much broad, methods are undefined and often limitless.

In a nutshell, your red team should evaluate your AI system on three key parameters:

- The severity of the consequences of a failure vector
- The probability of occurrence as found
- The likelihood of early detection of failure

The red team is functional, what next?

A functional red team is not just about finding the holes in your AI system. It is also about providing complete guidance and playbook to improve those weak points, plug those holes, and strengthen the system along the way, continuously.

Moreover, an effective red team operation wouldn't end after they have found a weakness in the system. It is just the beginning. The next role of the red team is to provide remediation assistance and re-testing. Also, more importantly, keep doing this as long as necessary.

There may be significant work involved in comprehending the findings, their impact, likelihood, criticality, and detectability and, furthermore, carrying out suggested remediations, retraining of machine with new data, and whole a lot, before your blue team says they're ready for the next round of testing.

The whole process of the red team finding weaknesses and blue team fixing them has to be an ongoing process with regular checks and balances. Avoid the temptation to do it once for the sake of it. Instead, make sure that you do it regularly and consistently. Doing so will help you to maintain a watch on the risk score of each aspect and monitor how you are progressing with the already established risk mitigation plan. Your target for each risk item in the list should be to reduce its risk score near zero.

You stress testing your AI system for any vulnerabilities is better than someone else exploiting it.

Handling Residual Risks

I often insist that when solving any problem, three things should be covered.

First of all, we must make sure that we are working on the right problem, that is, do the right thing. Without doing so, no matter what, we are going in the wrong direction.

Secondly, we must make sure that the solution to that problem is being developed right, that is, doing the thing right.

Thirdly, and most importantly, covering all your bases is a must. You may have the right solution for solving the right problem. However, if you miss the rest of the factors to solution success, your solution could be futile.

Doing everything you can is the best way to cover the risks. However, no matter how much you cover for, there still might be some lurking risk, which is either unknown or uncontrollable. This risk needs a different way of management.

A. Tamboli, *Keeping Your AI Under Control*,
https://doi.org/10.1007/978-1-4842-5467-7_10

What is a residual risk?

Residual risk in the case of an AI solution or implementation would be the amount of risk remaining after you taking care of inherent risks by way of risk controls. By performing a thorough pre-mortem analysis, ensuring that training and testing are complete, and, finally, systems have been stress tested using the red teams, you would be almost confident about the performance of the system and outcomes it shall achieve.

However, there still might be some risk that is unidentified or identified but can't be anticipated or controlled. Such residual risk can be dealt with by way of transference.

Transference or the risk transfer is a risk management and control strategy that involves the contractual shifting of a pure risk from one party to another. One example is the purchase of an insurance policy, by which a specified risk of loss is passed from the policyholder to the insurer.[1]

Existing options and limitations

If you have an IT solution, which is the closest equivalent for an AI solution, a few options are already available in the market for transferring some of the risks.

Information Technology Liability (IT Liability) insurance covers claims arising from the failure of information technology products, services, and/or advice.[2]

The information technology (IT) industry has unique liability exposures due to the crossover between the provision of professional services and supply of goods. Moreover, many service providers in this industry have a mix of both. It gets further complicated by the legal ambiguity around software advice and development and whether it is, in fact, the provision of a service or the sale of goods.

Traditional *Professional Indemnity* insurance policies often have onerous exclusions relating to the supply of goods, whereas traditional *Public and Products Liability* policies often contain exclusions relating to the provision of professional services.

Many insurers have developed a range of insurance options to address these issues, which they commonly refer to as *IT Liability* policies. These policies represent a combination of *Professional Indemnity* and *Public and Products Liability* insurances bundled into one product. These policies were developed to minimize the prospect of an uninsured claim due to its "*falling between the gaps*" between the two traditional insurance products.

---

[1] www.cna.com/web/wcm/connect/b7bacbf0-b432-4e0c-97fa-ce8730b329d5/RC_Guide_RiskTransferStrategytoHelpProtectYou+Business_CNA.pdf?MOD=AJPERES
[2] www.bric.com.au/information-technology-liability-insurance.html

However, the added complexity of AI solutions is driven by complex algorithms in addition to the data ingested or processed. Over time, change in data can significantly change product or solution characteristics; add to that cloud-based AI/ML services. It just widens the gap in existing options, and with that, the prospect of *falling through the gaps* increases significantly.

Is there a need for AI insurance?

Before we even think about AI insurance, I believe we must establish the need for it. And this need can only become evident when there are multiple issues with significant complexity.

There has been a steady stream of warnings since the last half-century to slow down and ensure we keep machines on a tight leash.[3]

Critical questions have been asked, such as who accepts the responsibility when AI goes wrong and what are the implications for the insurance industry?

Autonomous or driverless cars are the most important considerations for the insurance industry. In June 2016, a British insurance company *Adrian Flux* started to offer the first policy specifically geared toward autonomous and partly automated vehicles.[4] This policy covers typical car insurance options, such as damage, fire, and theft. Additionally it also covers *accidents specific to AI*—loss or damage as a result of malfunctions in the car's driverless systems, interference from hackers, failure to install vehicle software updates and security patches, satellite failure or outages affecting navigation systems, or failure of the manufacturer's vehicle operating system or other authorized software, the article stated.

Volvo has said that when one of its vehicles is in autonomous mode, Volvo is responsible for what happens.[5]

I think this is an important step. However, still, it fails to answer the question of who is liable for any accidents? Who is at fault if the car malfunctions and runs over someone?

When autonomous machinery goes wrong in a factory and disrupts the production, who is responsible? Is it the human operator who has thin-threaded control, or is it the management for buying the wrong system? Maybe it should be the manufacturer for not testing the autonomous machinery thoroughly enough.

---

[3] https://channels.theinnovationenterprise.com/articles/paying-when-ai-goes-wrong
[4] www.theguardian.com/business/2016/jun/07/uk-driverless-car-insurance-policy-adrian-flux
[5] https://fortune.com/2015/10/07/volvo-liability-self-driving-cars/

We need to establish specific protections for potential victims of AI-related incidents, whether they are businesses or individuals, to give them confidence that they will have legal recourse if something goes wrong.

The most critical question from a customer's standpoint would be, who foots the bill when a robot or an intelligent AI system makes a mistake, causes an accident or damage, or becomes corrupted? The manufacturer, developer, the person controlling it, or the system itself? Or is it a matter of allocating and apportioning risk and liability?[6]

Drew Turney, a journalist, argues in one of his articles, "We don't put the parents of murderers or embezzlers in jail. We assume everyone is responsible for his or her decisions based on the experience, memory, self-awareness, and free will accumulated throughout their lives."[7]

There are many examples where complex situations have occurred, which begets a need for AI insurance.

AI loses an investor's fortune

Early on, Austria-based AI company 42.cx developed a supercomputer named K1. It would comb through online sources like real-time news and social media to gauge investor sentiment and make predictions on US stock futures.[8] Based on data gathered and its analysis (based on that data), it would then send instructions to a broker to execute trades, adjusting its strategy over time based on what it had learned.

In May 2019, a Hong Kong real estate tycoon Samathur Li Kin-kan decided to sue the company that used trade-executing AIs to manage his account, causing him to lose millions of dollars. A first-of-its-kind court case opened up that could help determine who should be held responsible when an AI stuffs up.[9]

While it is the first known instance of humans going to court over investment losses triggered by autonomous machines, it also highlights the black-box problem of AI vividly. If people do not know how the AI is making decisions, who's responsible when things go wrong?

The legal battle is a sign of what's coming as AI gets incorporated into all facets of life, from self-driving cars to virtual assistants.

---

[6] www.theaustralian.com.au/business/technology/who-is-liable-when-robots-and-ai-get-it-wrong/news-story/c58d5dbb37ae396f7dc68b152ec479b9
[7] https://bluenotes.anz.com/posts/2017/12/LONGREAD-whos-liable-if-AI-goes-wrong
[8] www.insurancejournal.com/news/national/2019/05/07/525762.htm/
[9] www.bloomberg.com/news/articles/2019-05-06/who-to-sue-when-a-robot-loses-your-fortune

Karishma Paroha, a London-based lawyer, who specializes in product liability, has an interesting view. She says, "What happens when autonomous chatbots are used by companies to sell products to customers? Even suing the salesperson may not be possible. Misrepresentation is about what a person said to you. What happens when we're not being sold to by a human?"

Risky digital assistant for patients

In mid-2019, a large consulting firm started deploying voice-controlled digital assistants for hospital patients. The idea was, a digital device would be attached to the TV screen in the patient room, and the patient can request for assistance.

The firm wanted to replace the age-old call button service due to its inherent limitations. One of the significant limitations cited was that there is not enough context available for nurses to prioritize patients only based on the call request. It is possible, with the age-old call button system, that two patients have requested help, and one of them needs urgent help while others can wait. However, with just the indication of help request, nurses can't determine who needs immediate help. With voice-based digital assistants, the firm and hospital anticipated that with more context from the command text, they could prioritize nurse visits.

The system was deployed with prioritization based on the words uttered by the patients. For example, if the patient asked for drinking water, that request was flagged as a low priority, whereas if someone complained about pain in the chest, they were flagged with the highest priority. Various other utterances were prioritized accordingly. The idea behind these assigned priorities was, patient needing water to drink can wait for a few minutes, whereas a patient with chest pain may be at risk of attack and needs immediate assistance. Generally speaking, this logic would work in most of the scenarios.

However, this system wasn't linked with the patient information database. Such that, besides getting the room number of the requester, the system did not know anything else. Most importantly, it did not understand what the patient's condition was and why they were in the hospital.

As you would notice, not knowing the patient's condition or ailment may prove to be a severe impediment in some cases. These cases may be at the long tail of the common scenarios, but I think that is what makes it more dangerous.

For example, if a patient has requested for water, the system would put it as a low priority request and give other requests a relatively higher priority. However, what if the patient was not feeling well and hence requested water. Maybe not getting water in time could worsen their condition. It may be that they were choking or coughing and therefore asked for water. Continuous coughing may be an indicator of imminent danger. Without knowing the patient's ailment and condition, it is tough to determine whether a simple request for water is high priority or low priority; and, these types of scenarios are where I see significant risks. One may term this system as well-intentioned but poorly implemented.

Now the question is—in a scenario, where a high priority request gets flagged as a low priority on account of the system's limited information access, who is responsible? If the patient's condition worsens due to this lapse, who would they hold accountable?

Who (or what) can be held accountable for such failures of service or performance is a major lurking question; and perhaps, AI insurance may be able to cover the end users in all such scenarios when the need for compensation arises?

Regulations are gearing up

Since 2018, European lawmakers, legal experts, and manufacturers have been locked in a high-stakes debate: whether it's the machines or human beings who should bear ultimate responsibility for their actions.

This debate refers to a paragraph of text that was buried deep in a European Parliament report from early 2017. It suggests that self-learning robots could be granted "electronic personalities." This status could allow robots to be insured individually and be held liable for damages if they go rogue and start hurting people or damaging property.

One of the paragraphs[10] says, "The European Parliament calls on the commission, when carrying out an impact assessment of its future legislative instrument, to explore, analyze and consider the implications of all possible legal solutions, such as—establishing a compulsory insurance scheme where relevant and necessary for specific categories of robots whereby, similarly to what already happens with cars, producers, or owners of robots would be required to take out insurance cover for the damage potentially caused by their robots."

Challenges for AI insurance

Although the market is tilting quickly to justify AI insurance products, there are still a few business challenges. These challenges are clear roadblocks for any one player to take the initiative and lead the way.

---

[10]www.europarl.europa.eu/doceo/document/A-8-2017-0005_EN.html?redirect

A common pool

As the insurance fundamentally works with common pool, the first challenge appears to be not having enough quorum to start this pool. Several AI solutions companies or adopters can come together and break this barrier by contributing to the common pool.

Equitability

The second challenge is this common pool being equitable enough. Due to the nature of AI solutions and customer base, not every company would have equivalent revenue and pose a comparable risk. Being equitable, though may not be mandatory in the beginning, it will soon become an impediment for growth, if not appropriately managed.

Insurability of cloud-based AI

In the case of minors (children), parents are liable and responsible for their behaviors in the public domain. Any wrongdoing by them, parents pay for it. However, as they grow, and are termed as adults, responsibility entirely shifts to them.

Similarly, the liability of AI going wrong will have to shift from the vendor to the users over time, which may impact the annual assessment of premiums. Any updates to the AI software may bring this a few steps backward for vendors, as there are new inputs to AI now. If AI works continuously without any updates, the liability will keep shifting gradually toward end users.

However, in terms of cloud AI (cloud-based AI solution), this shift may not happen at all since vendors would always be in full control.

If customers or AI users supply the training data all the time, then there would be shared liability from solution and outcome perspective.

Attribution

However, of all, attribution of failure might be the biggest challenge with AI insurance. Several cases discussed throughout this book have shown us, how challenging and ambiguous it can be to ascertain fault contributing factor in the entire AI value chain.

AI typically uses training data to make decisions. When a customer buys an algorithm from one company but uses their training data or buys it from a different company and it doesn't work as expected, how much at fault is the AI and how much is the training data?

Without solving the attribution problem, insurance proposition may not be possible.

While writing this book, I met several insurance industry experts, who equivocally said one thing—a good insurance business model demands correct risk profile and history. This history doesn't exist yet. The issue is that it won't exist at all if no one ever takes the initiative and makes the flywheel rotate. The question, though is—who will do it first?

What might make it work?

While doing it first in the tech industry is quite a norm, in risk-averse sectors like the financial industry, it is just the opposite. So, until that happens, there might be an intermediate alternative.

How about insurers covering not the total risk but only the damage control costs? For example, if something goes wrong with your AI system, and it causes your process to come to a halt, you will incur two financial burdens. One would be on account of revenue loss and others to fix the system. In this case, while the revenue loss can be significant, system fixing cost may be relatively lower, and insurance companies may start by offering only this part of the cover.

Insurers may also explore *Parametric Insurance* to cover some of the known or fixed cost issues.

Aggregators can combine the risks of a cluster of several companies matching specific criteria. They can cover part of those risks themselves and transfer partial risks to the insurers.

Either way, it is not a complete deadlock to get this going.

Why the fixation on AI insurance?

Implementing AI solution means you can do things at a much larger scale. If you have been producing x widgets per day without the use of AI, you may end up creating 100x widgets with AI.

This type of massive scale also means, when things fail, they would also fail magnanimously. The issues 100x failure can bring upon a business could be outrageous. They typically not only result in revenue losses but also put a burden on fixing them, making alternate arrangements while they are being repaired, and so on.

In Chapter 2, I discussed a case study on Australian Telco's failed automation. If they have had AI insurance, perhaps things would have taken a different turn. The insurance approval process would have forced them to perform due diligence on their design and process up front, or else, they could have claimed for their losses under insurance. Either way, it would have been a good outcome for shareholders of the company.

This scale is dangerous, and therefore having an AI insurance would make sense. Additionally, with this option in place, it will also make people more responsible for developing and implementing AI solutions. These options would also contribute toward responsible AI design and use paradigm.

And more importantly, every human consumer or user would want some level of compensation at some point in time when AI solutions go wrong. They won't accept it if you say AI is at fault. It is a fair expectation to be compensated. The question is, who will foot the compensation bill?

And that is the biggest reason why I think AI insurance will be necessary. It may be a fuzzy concept for now but will soon be quite relevant.

Insurance or not, you are better off

Getting AI insurance, while can be a good idea, shouldn't be your objective. The primary objective of following a structured approach for AI development and deployment is to minimize risks to such an extent that it (AI solution) is safe, useful, and under control.

If you have followed three core principles of a good problem-solving, that is, doing the right thing, doing it right, and covering all your bases, then you would be better off.

Given that almost everything you would have mitigated or had planned to do so, there would be hardly anything that would qualify as residual risk. Managing residual risk is more like a remaining tail when the entire elephant has passed through the door.

Nonetheless, if there is uncertainty or unknown risk to your solution, risk transfer should be more effortless as the required due diligence would be done by you already.

It is always a good idea to deal with problems and risks in their smallest states. Stich in time saves nine!

When Working with Emerging Technologies

Throughout this book, I have attempted to explain the various possible risks related to AI solutions and ways to mitigate most of these risks or control them. Once these controls are in place, the world should be a better place. But as the technology adapts and changes, there could be new risks in the future, which are unpredictable at this point.

That brings me to the question I would like readers to ponder over—how can we pragmatically march toward a better future with AI in it? Should there be general guidelines for the widespread adoption of AI solutions in businesses and socioeconomical contexts? And if yes, what should these guidelines say?

I will attempt to strike a balance between optimism, pessimism, and skepticism while I elaborate on these points. But before I do that, let me begin with a story.

Several years ago, I was working with a home appliance manufacturing company. These were the early days of my career.

© Anand Tamboli 2019
A. Tamboli, *Keeping Your AI Under Control*,
https://doi.org/10.1007/978-1-4842-5467-7_11

My team and I would report to a senior manager, who was quite steadfast in pushing to have things exactly the way he wanted them. One day he asked us to prepare a report, where he asked us to emphasize key items of the report in italics and use colored fonts in a few places. He also wanted us to give him a print copy of this report.

The only (big) problem was, we were using a program called WordStar, and all we had was a dot-matrix printer! For those who do not know, it was a DOS (disk operating system) based word processing program. And, I do not have to explain why having a dot-matrix printer was a problem for the given task.

Our obvious response, in almost unison, was *not possible!* And he wouldn't listen. He kept pushing for us to make it happen. He insisted and said something on the lines of—don't try to fool me, I know that computer can do anything, and your boss promised that to us when buying it!

We all stared at our team leader, with somewhat mixed feelings, waiting for him to own up (after all, he promised the impossible) and fix it! Eventually, he did something behind the scenes that day, and we were saved.

That incident left a solid impression on me. This was not just the case of over promising of technological capabilities but also the one where business expectations were inappropriate.

Our interactions (with that senior manager) remained full of friction after that because we lost some credibility during this battle. Over promising and under delivering was undoubtedly one of the reasons.

Are we making the same mistake again?

Fast forward to today, almost 20 years have passed, and here we are—still dealing with these types of problems!

People are increasingly assuming that computers are better than humans and can do wonders. It feels like this notion has commandeered our intellect somehow. What is more, we still think that humans may get it wrong sometimes, but computers will not, ever!

This assumption is posing a different set of challenges for us. Moreover, these challenges are aggravating steadily, as computers become more pervasive and take part in our daily life in many ways.

It does not take a genius to know that the computers don't have a brain of their own, let alone intelligence or conscience. It is the developer, who instructs and teaches the computer—what, how, and when to do something. If developers make a mistake and poorly design and develop the code, it will be a problem.

If developers do not test their development appropriately or system integrators have used subpar hardware, or if there were fundamental flaws in the understanding of user requirements, the computer would perform poorly!

The proliferation of junk solutions

With democratized technology tools, this is becoming a bigger problem every passing day. As more technologies are democratized, the ability to create junk (knowingly or unknowingly) is increasing manifold. While some are using their wisdom while creating sensible solutions with excellent quality and usefulness, many others are doing just the opposite.

Several solution providers now believe that having a cool technology at disposal means they can create anything. Quite often, someone would ask, "Can we do this?", and then go through with it, if it's possible. However, the critical question we should be asking is, "Should we do this?". If the answer is no, then no matter how possible and easy creating any solution is, it should not be done.

However, the creators of AI solutions frequently get carried away. Even customers appreciate many solutions at their face value without realizing the utility of those solutions. It is an oft case with new products and technologies; novelty sells!

But then eventually, what gets created is a pile of junk. By junk, I mean solutions that are either subpar or trivial solutions that add low to no value. I am sure that these solutions would get disavowed at some point in time. But be known that they have already made their way into the market anyway.

However, with AI, several organizations still face difficulties in identifying metrics and quantifying benefits. The reason is apparent; most of them are approaching it the other way around. They have solutions that are looking for problems to solve.

Many businesses, I mean senior (responsible) managers in those businesses, still feel and believe that AI and automation will solve their problems for good. But when the technology is implemented, they find themselves in the soup again, just that it is differently flavored this time. This is a classic case when we work with junk solutions that do not solve real problems or do not focus on solving real problems.

The impact of low yield or no yield solutions can spread wider than you can imagine. By the time you implement and realize that the solution is not good, you already could be at the point of no return. Turning away could mean wasting too much money and resources; however, going further would also suggest you accept the subpar outcome as it won't get any better over time.

Businesses should start by raising expectations early in the adoption process. Be curious when the person in charge of such initiatives demonstrates only one side of the story. They usually do not tell the other side of the story, either because they do not know (lack of full knowledge) or they have some particular interest in doing so (often happens with vendors of technology).

Accepting the fact that *we don't know what we don't know* is quite critical here. But more importantly, if the issue is lack of knowledge, it must be handled before the next steps are taken. Without proper knowledge and information, it is nothing short of a gamble.

A good approach

There is a lot to talk about teaching machines morality, ethics, working in gray areas, and so on. However, we cannot wait for all the lights to turn green, and we must keep moving forward, learning and improvising as we do.

Machines make mistakes, just like humans do, and they will keep making them in future too. Businesses must accept this fact and know that machines also need, much like humans, attention, retraining, and performance improvement plan before they go live again.

Businesses must make sure, just as they do for humans, machines are progressively trained and rigorously tested before giving them more responsibilities. Any failure of a machines' performance should be dealt with, somewhat relatively stronger than humans.

If you have an HR team in your organization, make sure you augment them to handle additional resources that are not human. Create a HAIR (*Human and Artificial Intelligent Resources*) department. Appropriate policies need to be developed for managing these AI resources or digital twins of existing employees. This may sound a bit silly for now, but the direction we are heading toward would soon dictate that. A movement toward making AI transparent is catching up, and this could be an essential step for it.

However, there are many more things we can do to maintain sanity.

We must remain skeptical

Technology adoption is a relatively slow process and goes through several stages before being mainstream or being established as a norm.

Minimum viable product (MVP) is one such a stage through which several AI solutions are going through. For the established solutions, there is still a stage where user interaction with technology needs to be tested and customized. While the MVP is key to developing a successful product, treating MVP as MVP is very important.

The problem starts when someone sells their work, which could be still in progress, or has not achieved an acceptable working level yet! Their solution not working is less of a problem than when it starts working against the goals of a project.

We must remain positively skeptical, think through it sensibly, and adopt the technology with a grain of salt.

We must remember and acknowledge that the issue with many AI or automation solutions is not that they would break. Instead, complex solutions like these create leakage of some sort, which is not entirely wrong but not quite right either. This type of gray output piles over time, before one can realize that they own a heap of junk. Getting rid of that junk or fixing the completely flawed system could become a nightmare.

Remember that creating the MVP is excellent and essential, but we should deal with solutions that are not ready for full live deployment with utmost care and skepticism. Even if they are fully ready and validated, remaining skeptical will only work in your favor.

We must not overspeed

I say this because the problem is not the emerging technology and the bright future it promises. Our adamant belief that there is a bright technological future just around the corner is the issue!

So far, our attitude toward technology has been asymmetric and imbalanced. While we are reaping the benefits of the technology in this era, we are pushing the harm to our future generations in the form of polarization, lifelong unemployment, and concentration of wealth and power in the hands of the few technocrats. We are unfortunately racing toward something potentially bleak and unlikeable.

It could be that this is not the case; so far, we cannot tell. However, indeed, over-speeding kills, on roads or otherwise. We must take some time to think and reflect on repercussions and then build safety nets for negative impacts before we leap.

We must insist on a kill switch

Contemporary solutions are more than packaged software these days. Whether or not they are sophisticated AIs, they are still intricate pieces of intertwined technologies.

Older ways of ensuring that these solutions are working as expected are not suitable anymore. Quality assurance methods such as unit testing, regression testing, and so on are not enough to qualify emerging technological solutions, AI in particular.

Therefore, I recommend taking a systematic, thorough, and nontraditional approach in performing these tests. And I also insist that there should be a kill switch. Yes, there should be one kill switch, which may not be a physical switch per se, but some control mechanism, which can enable you, a human, to pull the plug and stop the system at once with no questions asked.

This is perhaps the most critical requirement you should state for each of your AI deployments. If there is no kill switch, you could be in trouble. Remember the debacle I referred earlier in Part I of the book for an automation bot? The company lost close to $12 million, simply because, before fixing the auto-bot for good, they were unable to switch it off, at least not without having a significant monetary and operational impact. It was a clear case of a poorly planned project. But it doesn't have to be the case every single time.

Ask yourself, or to your vendors of AI technology, "If I have to reverse everything that we are deploying now, how easy will that be?". If the answer to this question is anything other than a simple procedure, step back. You could be getting into a complex maze, never to return. As far as business risks are concerned, that should be a complete no-no situation. Insist on a kill switch and a simplified yet reversible procedure for the same.

Another critical question you must have an answer for—what is the alternative action plan when that (kill) switch is used, and automation or AI is shut down? In the earlier example I mentioned, there was no alternative, and there was no path of return, which created a big problem. Let us not repeat the same mistake twice. You don't have to live that mistake to learn either.

Additionally, make sure that you test the technology for real-life scenarios and evaluate possible and impossible (scenarios) both with an equal level of details. Then thoroughly audit the system for what it has learned, how and from whom, and then verify how that can affect you in the future. Do not just take creators' word for it.

Most importantly—pull the plug if needed, it is too costly to keep going with bad quality automation and AI than to just shut it.

We must demand quality

There is absolutely no reason to appreciate quantity over quality. I think it often promotes false economics, and here is the reason why.

Merely pushing for more quantity (of tech solutions) without paying attention to quality (of the solutions) adds a lot of garbage out in the field.

This garbage, when out there in the market, creates disillusionment and, in turn, adds to the resistance in healthy adoption.

This approach (quantity vs. quality) also encourages the bad practices by creators of AI at customer's expenses, where subpar solutions are pushed down customers' throats by leveraging their ignorance and an urge to give in to FOMO (fear of missing out).

Stuff like this happens during the hype cycle for any new/upcoming trend or technology. If this is happening with AI, it's nothing new per se, but we must demand quality and stop this from happening.

With AI, the quality problems are a bit tougher to gauge than usual. This is because the flaws in subpar techs do not surface until it is too late! In many of the cases, the damage is already done and would be irreversible. Unlike many other technologies, where things work or do not work, there is a significant gray area which can change its shades over a period. Moreover, if we are not clear in which direction would that be, we end up creating a junk.

We not only need to challenge the quality of every solution but also improvise our approach toward the emerging technologies in general.

This problem of lower quality often increases with unrealistic expectations from the technology. Additionally, failing to define the acceptance criteria up front can make it worse. Failing to establish such a criterion up front only results in the endowment effect. Implementation teams try to justify (much later) that whatever has been developed should be consumed because they worked on it so hard; worse part is when they try to retrofit the acceptance criteria, only to make it happen!

So necessarily, businesses do not have stubborn goals (read acceptance criteria) and do not have a handle on the means. This results in a somewhat uncontrollable situation. Development teams may tell you that the machine will learn eventually—but won't tell you when and how would it improve.

Garbage in will always result in garbage out—no matter how many years you keep doing it and how intelligent the machine is!

We must seek transparency

You might notice that people who make things often replicate their inner self in the output. A casual approach or thinking would encourage a casual output, and so would the precise thinking result in better output. Programmers who have clarity of mind and cleanliness in behavior usually create a clean and flawless program.

We must make people responsible and accountable for their output. In the case of an AI, if it does a lousy job, ask creators to explain and seek transparency. Hiding behind deep learning algorithms and saying, *"AI did it!"* cannot be an acceptable answer anymore.

People will often tell you that knowing inner mechanics is not essential for the users; only the output is. We must challenge this argument. For many other technologies, such as a car or microwave oven, this argument might hold—but for AI, it does not!

For each action automation or AI takes, creators should be able to explain why their technology did that! The reasoning should be completely transparent and interpretable when needed.

We must seek interpretability

This topic is being picked up quite frequently now, and there is some thought given to AI interpretability now. In general, *interpretability* means being able to explain to the end users why AI solution did something as an outcome. It should be not only able to explain why a particular result was derived but also should be able to explain how it arrived at that outcome.

This, in standard terms of six-sigma terminology, would mean, being able to explain:

$$y = f(x)$$

where **y** is the outcome (or outcomes) and **x** is the driver (or multiple drivers) of that outcome and *f* represents the process or formula that makes **x** turn into **y**.

Interpretability is important for many stakeholders of AI solutions. Primarily, for the end users, it is obvious to be wanting to know *why* a particular decision was taken, mainly because it directly affects them. Additionally, end users would also want to know that this decision has been made fairly and without any prejudice or bias. And most importantly, it is a must-have requirement for building trust.

Even from the creator's perspective, interpretability is quite useful. Mainly because it would only benefit them to build or fine-tune their models and algorithms.

There is a higher likelihood of being explainable AI solution if the solution is interpretable. Although it may not be the case always, it would be a good start, which is why businesses must seek interpretability in solutions and insist on this aspect for fair and trustworthy outcomes. Without interpretability, you will only keep facing resistance repeatedly while you lose trust in the eyes of end users.

We must seek simplicity

Whether it is traditional automation or a sophisticated AI program, all these necessarily do make decisions. And just like humans, more inputs mean more resources would be consumed and higher chances of errors and thereby lesser accuracy.

If it is a simple line or curve fitting algorithm, just a couple of points do not help. Similarly, a thousand points are also overkilling. There is an optimal limit at which maximum efficiency is achieved. We need to get to that point without wasting a lot of resources or time, and that is where simplification becomes relevant. As much as we do that for machine learning and computer algorithms, we should do the same for real-life decision-making too.

Simplification is also relevant from the algorithm auditing perspective. By audit, I do not mean a traditional regulatory audit but the review of the algorithm's work in general. If the work is not simple, the analysis can be complicated, lengthy, and sometimes misguided, and it may not yield proper results. Simplicity helps in retrospection process too.

If we insist on solutions to be simpler to audit and understand, transparent enough to be believed and accepted, we will have better control over the technologies.

Simple can be harder than complex. You have to work hard to get your thinking clean to make it simple. But it's worth it in the end because once you get there, you can move mountains.

—Steve Jobs[1]

We must seek accountability

As technology evolves and becomes more affordable, people are relying on it for their business-critical or life-critical functions. Bringing in people who would take accountability of outcomes seriously is often a better decision. Unless the application is for fancy or sheer playfulness, we need to put a responsible adult in charge.

---

[1] BusinessWeek, May 25, 1998.

Remember that machine learning is mainly teaching a computer to label things with examples rather than with programmed instructions. Besides, an AI that is based on this learning can do wrong things only in two circumstances: if the fundamental logic of decision-making programmed in it (by humans) is flawed, that is, lousy programming, or the dataset it was trained on is incorrectly labeled (by humans), that is, of bad data quality. In either case, the machine will *do the wrong thing* or *do it wrong*!

Here AI creators cannot just wash off their hands and say they do not know what went wrong when something goes awry. Humans need to take responsibility for embedding those wrong fundamentals and take accountability of outcomes.

The world, even today, looks at parents if their child does something wrong, questioning their upbringing methods and values. Then why cannot we see the same way to creators of AI solutions?

We should start assigning liability to creators of AI, much as we do now with guns or similar powerful things and make quality acceptance parameters stricter.

We must educate everyone involved

I keep saying this that *people often leverage your ignorance to sell you something*!

Getting educated on emerging technologies is quite essential to avoid these miss-sells. Starting from the top is even more critical. Executives should understand better as to what they are getting into.

They need to invest some time in acquiring this education to be able to ask the right questions at the right time/steps. Being more involved is necessary; merely learning some tech lingo is not going to be enough. If you are not involved, nothing else can fix it, ever!

Additionally, we may see that the same solution used by two different users can yield different results, much like two people can have the same smartphone with the same app, but still different usage or outcome.

Therefore, educating users and creators about technological impacts and interaction outcomes is the key. AI is not the kind of technology that will work the same way for everyone or every user. Human-computer interaction will be unique for each case, and that means the results delivered or the way they are provided can be different.

We must take user feedback seriously

Unfortunately, there is a general absence of experience-based knowledge in the developer community, so we cannot rely solely on the data and algorithms to know the right problems to solve. The same goes for business to customer or consumer interaction. This is a significant issue these days, and the neutral business perspective is often required to see things from objectivity lens.

My mention here of feedback loop does not relate to machine learning or AI system as such; it instead refers to the development and use value chain. When users provide feedback to creators and creators learn from it to adjust their creation, feedback serves better.

If creators consume feedback but never consider or act on it, it remains an open loop system. The open loop system is not stable.

Creators and users should ensure that an active feedback loop is established and everyone in that loop is sincerely learning from it.

We must prefer preemption over a fix

There are several solutions for each type of a problem in the market, even now, in such a nascent stage of AI. Indeed, not all the solutions could be a good match. Sometimes, their offerings are slightly different, and each use case may have different requirements too.

Ask yourself, if AI is such a great thing, why should we focus and limit its applications in solving problems? Why don't we preempt issues and avoid them altogether, won't that be a wise thing to do? How many technologies stop issues from becoming problems today?

Some solutions can provide a fix to the problem as an after the fact (e.g., automated fixing of a broken machine), while some would give an early indication (e.g., anomaly detection in machine performance). Some will help in improving problem-fixing performance while some will help in identifying problems early on.

Focus and prefer preemptive systems than fixers. Both are important, but if you must—prevention is always better than fix!

Working with AI responsibly

As a common sense, everyone should take responsibility for their business. Whether one is creating an AI tool or using it. Remember that when you create a tool, it can change the way business is done, it can change culture and society profoundly. So, creators of the tool become responsible for all those consequences eventually. Just like the factory owners are responsible for environmental pollutions or emissions and waste, AI creators would be accountable for their solutions.

And almost equally, business users of AI would be responsible too, first for encouraging creators for developing it and secondly for consulting them with inputs.

In his book, *Technology vs. Humanity,*[2] Gerd Leonhard says, *"Embrace technology but don't become it. Radical human augmentation via technological means will be a downgrade, not an upgrade."* We must embrace AI in a way that is appropriate and just enough.

We all know that technology does not have ethics, but our society depends on them. Which is why just because we can do something does not mean we should do it. The responsible approach is the central requirement to work and handle powerful technology like AI.

Sanity is the key!

Sometimes, it is better to deal with humans than machines. When to do that is a rational thought process, where sanity is necessary!

We should not get carried away and assume that just because we have smart technology, we can use it to solve every problem around us. AI is much like a new hammer; let us avoid treating all of the issues as nails and avoid rushing into the emerging future. It is challenging and sometimes nearly impossible to undo strategic and technological mistakes these days.

When working with emerging technologies, sanity is the key!

---

2 Fast Future Publishing, 2016.

The Leash System

AI is a powerful technology with limitless applications; and, with great power comes great responsibility!

However, while everyone knows *why* responsible AI is necessary, not many know exactly *how* to be responsible with AI. To support this aspect, I have developed a structured methodology based on the principles mentioned in this book. This methodology is known as *The Leash System*. It is a prescriptive recipe that can be tweaked same as any other recipes (Figure A-1).

© Anand Tamboli 2019
A. Tamboli, *Keeping Your AI Under Control,*
https://doi.org/10.1007/978-1-4842-5467-7

Figure A-1. The Leash System

The Leash System is general enough to be applicable for the majority of the AI implementations. It is comprehensive and covers all aspects of the design, development, and deployment. Most importantly, it is actionable, which enables you to take steps to make *responsible* AI a reality.

Learn more about **The Leash System,** download tool templates explained in this book, and engage in a conversation here: https://leash.knewron.app.

Index

© Anand Tamboli 2019

A. Tamboli, *Keeping Your AI Under Control*,

https://doi.org/10.1007/978-1-4842-5467-7